# Carbon Yoga:
## The Vegan Metamorphosis

We are in the midst of a monumental transformation in human civilization, akin to a metamorphosis. Just as in Nature where the caterpillar gorges itself before forcibly undergoing a metamorphosis in the chrysalis, we over-consuming humans are reaching a point of being forcibly transformed within the cocoon of our finite planet into compassionate, life-affirming butterflies. *Vegan* life-affirming butterflies.

*Carbon Yoga* weaves a narrative through the events of the past to show that such a vegan metamorphosis is inexorable and that this system change will occur within the next decade, both from a scientific perspective and from a religious and spiritual perspective. The narrative proposes an evolutionary purpose for humanity and for Western civilization. During the Caterpillar phase, humans unintentionally built all the tools and technologies needed to regulate the Earth's climate and organize an equitable global human society. But as a consequence, the Earth is marinating in ever accumulating toxic pollution even as ecosystems are degraded and the climate is changing. Now, we have no choice but to treat all Life as sacred in order to preserve that on which we depend for our own survival. Such a revolutionary change in human outlook has the added benefit of inducing spiritual awakening, resolving social justice issues, elevating our connection with all Life and improving human health and well being. It is only such a quest for "moral singularity" that will heal the Earth, regenerate Life and lead us to true global sustainability.

*Carbon Yoga* is the second in a series that began with the 2011 book, *Carbon Dharma: The Occupation of Butterflies*. While *Carbon Dharma* was mainly about what we need to do, as a species, to reach sustainability, *Carbon Yoga* takes this understanding of ethical and environmental righteousness (*Dharma*) and suggests how we can go about doing it (*Yoga*).

A Climate Healers Publication

ISBN-13: 978-1533019295
ISBN-10: 1533019290

Email: info@climatehealers.org
Publisher Website: http://www.climatehealers.org
Book Website: http://www.carbonyoga.org

Front Cover Photo: Illustration of the "Moral Singularity" by the mouth painting artist, Srilekha Mandalapalli. The flute represents the "Kundalini" with the seven holes corresponding to the seven chakras that have to be opened to reach a state of enlightenment, The peacock feather symbolizes freedom from worldly attachments. More such amazing paintings by Foot and Mouth artists can be found at http://www.imfpa.org

Foreword: Courtesy Dr. Will Tuttle.

# Carbon Yoga:

## The Vegan Metamorphosis

## Sailesh Rao
with Foreword by Dr. Will Tuttle

A Climate Healers Publication

*All proceeds from the sale of this book goes to support the work of Climate Healers.*

Sailesh Rao is the Founder and Executive Director of Climate Healers, a non-profit dedicated towards healing the Earth's climate. A systems specialist with a Ph. D. in Electrical Engineering from Stanford University, Stanford, CA, USA, conferred in 1986, Sailesh worked on the Internet communications infrastructure for twenty years after graduation. In 2006, he switched careers and became deeply immersed, full-time, in the spiritual and environmental crises affecting humanity. He is the author of the 2011 book, *Carbon Dharma: The Occupation of Butterflies*.

Foreword by Dr. Will Tuttle, author of the best-selling *The World Peace Diet*, a healthy, enthusiastic, and profoundly happy vegan since 1980, and Dharma Master in the Zen Buddhist tradition. Will sees his mission as one of bringing the message of radical inclusion to our wounded and fragmented culture. For this reason, he lectures, teaches, and performs throughout North America and worldwide. Because of this, he's a recipient of The Peace Abbey's "Courage of Conscience Award," as well as the "Empty Cages Award" and the "Shining World Hero Award."

# Table of Contents

For Kimaya and the children of the Miglets.

*"There are two most important days in your life: the day you were born and the day you discover why you were born"* - Boniface Mwangi.

# Foreword

It is becoming increasingly obvious that a new cultural narrative is required if we are going to be successful in our human quest for wisdom, meaning, and harmony in our rapidly changing world. Understanding how the deep structure of our culture confines us to the shallows of consciousness provides keys to this new narrative, and to authentic personal and planetary transformation. Learning to connect the dots between our culturally mandated mistreatment of animals and the environmental, cultural, psychological, and health challenges we face is an essential key to evolving new perspectives that provide the foundation and inspiration for lasting positive change.

Our confinement to the shallows of living is due primarily to the enormous power of the food program that is injected into all of us from infancy through our culture's meal rituals of disconnectedness that require us to participate in the abuse and commodification of living beings. The beckoning frontier of authentic positive change lies in understanding the consequences of these food actions and attitudes, and in bringing our lives into alignment with this understanding, at both the individual and collective levels. Veganism is thus increasingly recognized as the cutting edge of engaged and empowered social and environmental healing in our world.

The ideas in this book can guide and inspire us on this adventure of transformation. It's dawning on us all that fundamental changes in our orientation to our Earth, to animals, and to each other are required. We can see that continuing to function primarily as consumers is ravaging our beautiful and abundant planet in ways that destroy the viability of the living web that supports all life here. In this book's metaphor of the Caterpillar, we can see that there is nothing in the Caterpillar's existence or approach to life that would predict his sudden cessation of destructive consumerism and subsequent transformation, and the same is in many ways true of us. And yet, at a certain point, the Caterpillar simply stops his rampant consuming and turns his attention into a completely different direction, undergoing a complete metamorphosis and emerging through this as the beautiful, delicate Butterfly, dancing on breezes, sipping nectar, and inspiring us humans with the possibility that something similar could happen with us.

As the primary cross-cultural symbol of transformation, Butterflies exemplify experiences that many of us have had in our lives, and for many of us, going vegan is primary among these transformational experiences. As beings with an innate yearning to learn, grow, and create, and also with a basic sense of ethics, we are naturally interested in understanding how our lives and actions affect others. But because we are born into a culture that mandates relentlessly eating the flesh and secretions of abused animals, we disconnect from our inherent intelligence and empathy, and lose track of our purpose on this beautiful and abundant Earth. Now, we see signs everywhere that we are awakening from this culturally-imposed trance of rampant consuming and abuse, and beginning to discover our purpose as benefactors and protectors of life here. How all this will play out in the immediate several decades we are facing is fascinating and daunting to contemplate indeed.

May our hearts and minds open to deeper wisdom and compassion for each other and for all life in these critical times. And may we understand, as this present volume helps illumine, both the consequences of our actions and also the positive vision of a future that beckons to us. Perhaps, as this book suggests, we have been fulfilling our destiny all along, and now it is time to take the next evolutionary leap.

The old stories of human superiority—of domination, exploitation, and rationalization—are becoming loathsome in our mouths, and as we spit them out in disgust, sensing their toxicity, our yearning hearts discover new narratives that feed and satisfy us with their health, meaning, and nourishing beauty. These new stories draw on the ancient wisdom of the infinite interconnectedness of all expressions of life, and of learning to respect our intuitive yearning to evolve. As we make our effort to understand and contribute, we help humanity to awaken. We can discover and fulfill our purposes, and celebrate our lives and bless each other, as we are intended.

Dr. Will Tuttle

# Preface

*"The only myth that's going to be worth thinking about in the immediate future is one talking about the planet -- not this city, not these people, but the planet and everybody on it"* - Joseph Campbell.

I began writing this book over two years ago. I attended a Vipassana retreat[1] in Twenty Nine Palms, Southern California in January 2014 and during those ten days of silence and meditation, the patterns of this unabashedly positive story began to emerge. It was stunning! As a systems specialist, I'm now in total awe of the systems design that is Nature. I'm talking about Nature as a whole, including us humans, with all our blatant flaws. It all fits beautifully when we consider the crime scene backwards. As Sir Arthur Conan Doyle gets Sherlock Holmes to say in *A Study in Scarlet*[2],

> "In solving a problem of this sort, the grand thing is to be able to reason backwards. That is a very useful accomplishment, and a very easy one, but people do not practice it much."

I had not practiced it much either, but this was precisely what it took to thread this story. But before I begin, here's my background so that you understand my perspectives.

I was born to a Hindu Brahmin family in South India in 1960 in the village of Iddya near Mangalore and was raised and educated in Chennai, the fourth largest city in India. In 1981, I immigrated to the United States to pursue my graduate studies in Electrical Engineering.

Chennai recently suffered a "once in a thousand year flood" during the NorthEast monsoons, affecting millions of lives[3]. Climate change clearly loaded the dice in favor of such an extreme flood event occurring and this loading will only get worse if the Earth's temperature continues to rise. My birthplace, India, is expected to be one of the most affected countries due to climate change, biodiversity loss, desertification and toxic pollution. Therefore, I have a lot at stake, personally, to ensure that our human story rights itself without an apocalyptic loss of lives.

Prof. Thomas Kailath at Stanford University in California trained me as an electrical systems engineer, in the early 80s. Prof. Kailath is such a giant in this field that he has gone on to receive numerous international accolades, including the U.S. National Medal of Science from President Obama and the Padma Bhushan, the third highest civilian honor bestowed by the Government of India. He taught me that a genuine systems specialist must peer over the shoulders of experts in different fields of study in order to discern cross-disciplinary patterns. Thus, a systems specialist is mainly an integrator of the individual stories told by other storytellers in their respective fields of expertise. In order to be effective, he or she must have the curiosity to

> "Know something about everything and everything about something,"

as Prof. Donald Knuth, from Stanford University, put it recently[4]. About twenty years ago, I thought I knew "something" about every conceivable source of interference that could affect the quality of data communications over the Internet and "everything" about exchanging data over twisted-pair copper wires through silicon chips. That's how I became deeply involved in designing the hardware infrastructure of the Internet in the 90's. It was professionally rewarding as the Internet exploded in popularity!

Until about 10 years ago, I considered my work on the Internet to be some of the most difficult and interesting systems challenges in the world. Then, one evening, it changed. I happened to be watching former Vice President Al Gore's Global Warming slide show on LinkTV and it changed my life[5]!

The environmental systems problem that he described was more challenging than anything that I had ever encountered in my technical career. I realized how tremendously consequential it was for the legacy we were leaving our children. For the past ten years,

I've been working full time on environmental issues, while staying self-funded. In 2011, my book, *Carbon Dharma: The Occupation of Butterflies*[6], was published about "what" we need to do as a species to reach sustainability.

This book is a follow-up on "how" we can go about doing it. It is also an updated story of the patterns that I have discerned so far, five years later, in this mother of all systems challenges.

Just as the author and public theologian, Brian McLaren, had done in "*Everything Must Change*[7]" from a Christian perspective, this book connects our environmental and economic crises with the spiritual crises that humanity is facing. As was the case in *Carbon Dharma*, this book draws upon ancient Hindu texts extensively, mainly due to my familiarity with the Hindu tradition. However, in this book I've tried to include quotes from the Holy Bible and the Holy Quran to show that the Rig Vedic verse:

> "*Ekam Sat, Vipra Bahudha Vadanti* (The Truth is One; The Wise Call It by Many Names),"

holds when it comes to all long-standing faith and wisdom traditions. There should be no daylight between us in our response to our global predicament.

As a systems engineer, I'm not a trained philosopher or theologian, though I'm fascinated by all faith and wisdom traditions including Secular Humanism, and have been studying them all. I'm not a trained environmentalist or climate scientist, though I've been poring through the scientific literature on climate change and environmental degradation for the past ten years. I'm not a trained nutritionist or food scientist, though I've completed the online course on Plant-Based Nutrition[8] from the T. Colin Campbell Center for Nutrition Studies and eCornell and have actively experimented with my diet over the past 30 years. I'm not a trained economist, though I've been closely following the latest developments in financial markets and in digital currencies. I'm not a trained sociologist, though I have been immersed in social justice issues for many decades. I'm not a trained astro-biologist, though

I've been fascinated by the possibility of extraterrestrial life for as long as I can remember. I mention these varied disciplines because the story in this book draws upon these fields a fair bit. Thanks to the Internet, I had ready access to the accumulated knowledge in all these disciplines and to many generous subject matter experts who patiently led me through the nuances of their respective fields of study. However, I must hasten to add that while these varied sources provided the plausible "dots" for this story, any misinterpretations that have seeped through are really due to my limited understanding of their respective fields.

I've also had the pleasure of working alongside so many passionate people from Non Governmental Organizations (NGOs) such as the Foundation for Ecological Security[9] (FES) and SAI Sanctuary[10] in the villages of India and the Climate Reality Project[11] and the movement for American Indian rights[12] in the US. These generous activists and indigenous people that I met during the course of my field work at Climate Healers have truly contributed as much to my growth and understanding and to this story as all of my respected colleagues in the scientific realm.

The Booklist, in its review of Ian Morris's book, *Why the West Rules - For Now: The Patterns of History and What They Reveal About the Future*[13], wrote that[14],

> "Only the supremely self-confident put forth all-encompassing theories of world history, and Morris is one such daredevil."

Ian Morris, a historian and archaeologist, with a chaired position as Professor of Classics and Professor of History at Stanford University, is most certainly qualified to put forth such all-encompassing theories. This book does propose such an all-encompassing theory of world history, but it isn't supreme self confidence that compelled me to write it. Rather, it is my undying love for my children and their generation, and above all, my undying love for my granddaughter and her generation. Our future generations are simply amazing and deserve better than the apocalyptic stories that they are being told today. If our children don't have a positive story of their world to work with, then have we not failed as parents? Therefore, given my systems training, I felt that it is my duty to discern a positive story that fits the same underlying facts and reality, even as our current socioeconomic system grinds towards an inevitable collapse.

The system is breaking. There is widespread discontent throughout the world, especially among the youth. Ever since world leaders resolved the financial collapse of 2007-8 with massive bank bailouts, the youth of the world have been rebelling against the established order. Despite appearances to the contrary, the Greek movement of 2008 and the Tunisian, Occupy and Indignados movements of 2011 have remained potent. They have simply morphed into numerous social justice movements that are beginning to have a substantial impact on political processes throughout the world. In the United States, the Occupy movement has been followed by the DREAM'ers, the Fight for 15, Black Lives Matter, Direct Action Everywhere, Collectively Free and the Climate Mobilization movement, to name a few. They have propelled a rumpled, socialist candidate, Sen. Bernie Sanders, to within a hair's breadth of the nomination for the Presidency of the United States in the Democratic Party.

The system is breaking. Last year, I was invited by Rep. Raul Grijalva to speak on a panel about the Trans Pacific Partnership(TPP)[15], a trade deal so egregious that it is unimaginable that any conscious legislator could vote for it and hope to win reelection. The TPP contains every possible half-baked idea ever devised to maximize corporate profits and socialize corporate risks. It's as if the political establishment is terrified of corporate profit reductions triggering a system collapse and is going out of its way to ensure corporate well being.

The system is breaking. The Republican Party in the US is serving up Presidential candidates who no longer mince words about their racist, misogynist, bigoted, fascist policies. Unlike the Democratic Party, even the pretense of upholding "liberty and justice for all" is gone, as social divisions are laid bare and exploited by the leaders of this once Grand Old Party.

The system is breaking. Official secrets are leaking, not in dribs and drabs, but in gushing torrents, exposing the powerful and the ruthless games they play. It is now a good policy for every official to assume that anything they say or do will be exposed, if not now, then at some point in the future. Therefore, radical transparency is not only the best policy, but it is the default option as well.

To paraphrase the great Persian poet, Islamic scholar and Sufi mystic, Jalal ad-Din Muhammad Rumi, or more fondly, just plain Rumi, where the system is breaking, that's where the light is entering[16]! This is undoubtedly an optimistic view, but I am an optimist. I'm also an unabashed salesman of *Ahimsa*, the ancient Vedic doctrine of non-violence towards all Life. Many of my meat, fish and dairy consuming friends and relatives will tell you that I take every opportunity to advance this cause. Reasoning backwards, I was amazed to discover that the underlying connecting thread for the story of *Carbon Yoga* is indeed compassion for all Creation. It is the head telling us that only the heart has the solution! While I must hasten to add that Carbon Yoga is just one plausible view of all that has happened so far and the scenario I've outlined is just one possible path forward towards a sustainable future, as far as I can tell, the path of compassion seems to be the only way towards a positive future that doesn't involve genocidal interludes.

Finally, while writing this book, I've tried to quote directly from various sources so that the story gets integrated mostly in their original words. As much as possible, I've tried to be the faithful recorder of the story as it was being told all around me by countless others.

Sailesh Rao.
Phoenix, AZ.
April 2016

# 1. Our Stories Are Failing Us

*"The old is dying and the new cannot be born; in this interregnum, a great variety of morbid symptoms appear "* - Antonio Gramsci.

Everything in Nature is unique. Every human being, every animal, every leaf, every rock and even every grain of sand is unique. In turn, the universe experienced by every being is unique as well. Rumi's words[1],

"Stop acting so small. You are the universe in ecstatic motion,"

are literally true.

The universe you experience is yours and yours alone. It is this sheer complexity of Nature that inspires us to tell stories to make sense of reality. Stories help us discern patterns in Nature. The language and images we use to tell stories help simplify reality into bite-sized chunks that we can then use to recognize the commonality of our experiences.

Every story is unique. What anyone absorbs from any story is unique as well. The American writer of comic fantasy, Christopher Moore, wrote in his book, *Practical Demonkeeping*[2]:

"Everything is a story. What is there but stories? Stories are the only truth."

It's all just stories. While truth itself can only be experienced and cannot be expressed in words, communicating the truth requires us to tell stories. All religions expose deep truths in stories. The historian, Yuval Noah Harari, attributes the biological success of our species to our ability to construct common stories that we all believe in, that allow us to organize ourselves in far greater numbers than any other species and thereby dominate them[3]. We mentally construct these common stories by synthesizing individual stories into a social consensus. For the past 500 years or so, we have also been using the scientific method to establish some order in how we construct the common stories of our lived reality[4]. But of late, the common stories of our lived reality have been failing us.

## 1.1 The Story of Endless Growth

In 2003, the Nobel Laureate, Richard Smalley, compiled a Top Ten list of problems that humanity faces today[5]:

1. Energy
2. Water
3. Food
4. Environment
5. Poverty
6. Terrorism and War
7. Diseases
8. Education
9. Democracy
10. Population

Indeed, all of them are usually referred to as "crises" these days, the Energy crisis, the Water crisis, the Food crisis, and so on, down the list to the Population crisis. True to the implicit separation between science and religion that has persisted in the West for the past 500 years, the Spiritual crisis or the crisis of separation from Creation, which many believe to be the root of all these crises, is conspicuously absent in this list. Furthermore, the first three in Smalley's list, Energy, Water and Food, are really just escalating demands that we make upon Nature, not "problems." Our demands have been increasing exponentially to date due to global economic growth, and assuming that we continue on this exponential growth trajectory, the UN projects that humanity will require 50% more food calories, 45% more energy and 30% more fresh water by 2030[6]. But to put these demands in perspective, the average human being is currently harnessing the equivalent of 22 energy slaves, mostly fossil-fuel based, with the average American harnessing the equivalent of 150 energy slaves each! Yet our expectation is that even the average American is entitled to even more energy slaves, though preferably fueled through clean energy sources.

Except in fringe circles, there is scientific consensus that all the environmental crises in the world, climate change, biodiversity loss, desertification, toxic pollution, etc., can be largely attributed to these escalating human demands for energy, water and food. Scientific projections show that our current environmental trajectory cannot continue for much longer. For instance, in a comprehensive survey of over 10,000 species, the World Wildlife Fund (WWF) reported that the total biomass of all wild vertebrates decreased by 52% in the 40-year span between 1970 and 2010[7]. During that time, human population approximately doubled and human per capita consumption also approximately doubled so that human impact on the environment approximately quadrupled[8]. If such exponential growth in impact continues unchecked, we can mathematically show that ALL the wild vertebrates will disappear by the year 2026! In response to this carnage, the environmentalist and author, George Monbiot, wrote a gut-wrenching column with this plea in the Guardian[9]:

"If the news that in the past 40 years the world has lost over 50% of its vertebrate wildlife (mammals, birds, reptiles, amphibians and fish) fails to tell us that there is something wrong with the way we live, it's hard to imagine what could. Who believes that a social and economic system which has this effect is a healthy one? Who, contemplating this loss, could call it progress?

Is this not the point at which we shout stop? At which we use the extraordinary learning and expertise we have developed to change the way we organize ourselves, to contest and reverse the trends that have governed our relationship with the living planet for the past 2m years, and that are now destroying its remaining features at astonishing speed?

Is this not the point at which we challenge the inevitability of endless growth on a finite planet? If not now, when?"

Even the normally staid Forbes magazine is now sporting headlines like, *"Unless it Changes, Capitalism will Starve Humanity by*

*2050*[10]!" The accompanying Drew Hansen article contained stinging lines such as:

> "Capitalism has generated massive wealth for some, but it's devastated the planet and failed to improve human well-being at scale".

> "Corporate capitalism is committed to the relentless pursuit of growth, even if it ravages the planet and threatens human health".

Storytellers such as Paul Gilding, an activist and author of The Great Disruption[11], have noted that we are already too obese for our host planet and that we have no option but to scale back on the human impact on Earth and mature out of our adolescent, exponential growth phase. In his TED talk from 2012, Paul Gilding said[12],

> "The Earth is full. It's full of us, it's full of our stuff, full of our waste, full of our demands. Yes, we are a brilliant and creative species, but we've created a little too much stuff -- so much that our economy is now bigger than its host, our planet. This is not a philosophical statement, this is just science based in physics, chemistry and biology. There are many science-based analyses of this, but they all draw the same conclusion -- that we're living beyond our means. The eminent scientists of the Global Footprint Network, for example, calculate that we need about 1.5 Earths to sustain this economy. In other words, to keep operating at our current level, we need 50 percent more Earth than we've got. In financial terms, this would be like always spending 50 percent more than you earn, going further into debt every year. But of course, you can't borrow natural resources, so we're burning through our capital, or stealing from the future.

> When I say full, I mean really full -- well past any margin for error, well past any dispute about methodology. What this means is our economy is unsustainable. I'm not saying it's not nice or pleasant or that it's bad for polar bears or forests,

though it certainly is. What I'm saying is our approach is simply unsustainable. In other words, thanks to those pesky laws of physics, when things aren't sustainable, they stop. But that's not possible, you might think. We can't stop economic growth. Because that's what will stop: economic growth. It will stop because of the end of trade resources. It will stop because of the growing demand of us on all the resources, all the capacity, all the systems of the Earth, which is now having economic damage."

The fifth in Smalley's list, Poverty, is usually touted as the reason why we need to continue the exponential growth that is causing all this damage. But despite the exponential growth in our world economy over the past 40 years, there are over 3 billion people living in abject poverty today, on less than $2 per day, almost as many as the entire human population of the Earth in 1970[13]! Therefore, it is hard to see why our situation would improve 40 years from now, if we persist with that exact same growth strategy.

As for the sixth in Smalley's list, Terrorism is an asymmetric response to War, which seems to be fought nowadays to destroy the infrastructure of small nations so that corporations can rebuild them, again to foster economic growth[14]. Diseases are mostly self-inflicted through the social promotion of inappropriate and excessive consumption so that the symptoms can be corrected through pharmaceuticals[15]. Both of these fall under the rubric of "broken-windows" economic growth[16], but thus far, technological advances have helped us postpone the day of reckoning for such misguided policies to foster growth.

However, technologists are now weighing in with the news that Moore's law[17], the cornerstone of the exponential growth in human productivity over the past few decades, is saturating in many respects. Moore's law is attributed to Gordon Moore, the co-founder of Intel Corporation, as he predicted in 1965 that the number of transistors on a silicon chip would double every 18-24 months. Corollary laws were propounded that the clock speed and the power consumption of an Intel microprocessor would double every 18-24 months as well. And all these laws held true for 40 years.

Then the clock speed and power consumption of Intel Microprocessors saturated around 2005. While the number of transistors in Intel Microprocessors have continued to double every 24 months, they have increased mainly in the form of "dark" silicon, which are circuits that are turned off at all times except when special functions are executed. Even this "dark" doubling is scheduled to stop in the next few years as we hit physical limits[18].

## 1.2 The Story of Technology

Nature's physical limits are not easy to overcome. Certain laws of physics are literally sacrosanct as engineers have been butting their heads against them for decades without much success. During my technical career, I experienced the thrill of discovering engineering breakthroughs as well as the futility of overcoming Nature's physical limits.

The mid nineties were heady days in the history of the Internet. The Internet community was confined mainly to technical users then, but we were laying the foundations for the popular version of the Internet that emerged at the turn of the century. In those early days, the large transnational telecommunications corporations were huge supporters of a tiered data communications system called Asynchronous Transfer Mode (ATM), where all data transmission requests were to be sent to a central server, which would then determine precisely how the data gets routed from source to destination over the Internet. But the small Silicon Valley companies were promoting a peer-to-peer, distributed networking technology called Ethernet, where any computer server could plug in and participate as an equal in the routing of data, in keeping with the nascent principle of net neutrality of the Internet. As you can imagine, if the ATM protocol had won the data communications war then, the Internet would have been very different and John Oliver wouldn't be railing against the Federal Communications Commission (FCC) in 2014 as it tries to eradicate net neutrality[19].

Net neutrality would probably never have existed at all!

Indeed, by late 1995, the ATM Forum had standardized on a 155.52 Mb/s (million bits per second) data rate for its fastest links on copper wires, while the Ethernet standards committee had standardized on a slower 100 Mb/s data rate. Yet the freewheeling Ethernet protocol was being rapidly deployed in the marketplace due to its inherent simplicity and due to the robustness of its slower 10Mb/s copper links. But there was a hitch. The newer 100Mb/s Ethernet links were experiencing intermittent failures, causing server connections to break down and resulting in customer complaints. The chairman of the Ethernet committee at that time was worried enough to ask me, as a systems specialist, to take a look at the protocol and suggest ways to improve it.

The next committee meeting was to be held on Jan. 8, 1996 and I had a flight to catch in the morning of Sunday, Jan. 7 to make my presentation to the committee. Standards committee work is meant to be done *pro bono* and I normally gave it lower priority than my day-to-day work that paid our bills. Thus, as usual, I spent the evening before the flight at my office in New Jersey studying the problem and preparing for the presentation. To my incredible delight, I discovered that while the 100Mb/s protocol had some niggling issues in the specifications, the physical wiring itself was capable of easily supporting a data rate that was TEN times faster, i.e., 1000Mb/s or 1 Gigabit per second! I was so sure that the committee would be thrilled with that discovery!

Late that night, after finishing up my presentation, I turned off the lights in the office and pushed on the front door to go out. It wouldn't budge! It was completely blocked by wind drifts from the snow that had been falling through the evening and I was stuck in the office. Overnight, a major nor'easter dumped 14 inches of snow in our area and up to 48 inches in other places and the whole state of New Jersey was shut down. This was the first and only night that I slept on the couch at the office. Early next morning, my wife, Jaine, called the maintenance crew and had them clear a narrow pathway to our office door first so that I could get to Newark airport on time. I still recall the eerie feeling as I rushed between two huge, ten-foot tall walls of snow on either side of me in front of the office, to catch my flight to make my presentation the next day.

At the end of the presentation, everyone in the room laughed! The chairman of the committee said,

> "We're having trouble getting 100 Megabit to work and you are telling us that we can crank it up 10 times faster? Dream on! I'll believe it when I see it."

Nevertheless, he was kind enough to let the committee flesh out these ideas. It turned out that my calculation on that eerie snow-bound night was real, not a fantasy. A concrete proposal that I developed a few months later became a standards protocol.

The resulting 1000BASE-T standard was the first widely deployed communications standards protocol that relied on high-speed Digital Signal Processing (DSP) technology for Ethernet as opposed to the 100BASE-T standard that could be implemented with standard analog techniques[20]. 1000BASE-T was adopted as a standard in 1999 and during the heydays of the Internet mania, annual shipments were over 150 million units by 2003, within 4 years. The digital 1000BASE-T technology developed a reputation for being more robust on the same cabling than any analog 100BASE-T device and it still forms the backbone of the Internet infrastructure. The wired network connection on your laptop is most likely a 1000BASE-T Gigabit Ethernet link.

Starting in 2003, I began attending the 10 Gigabit Ethernet Task Force meetings, where a more sophisticated DSP technology, 10GBASE-T, was being discussed to propel data ten times faster than 1000BASE-T over similar copper wiring. The same chairman who had told me to "Dream on!" seven years earlier had now become a wide-eyed techno-optimist and he was clearly expecting 10GBASE-T to be a piece of cake based on his 1000BASE-T experience. However, this time around, we were reaching physical capacity limits on the wiring and I knew that it was not going to be easy at all. Nevertheless, the Ethernet standards committee approved the 10Gigabit Ethernet on Copper standard, 10GBASE-T, in 2006 almost unanimously. I stopped working on it, partly because I felt that it would not meet the high expectations of the user community, but mainly because I became interested in environmental issues instead. Five years after that standard was approved, the worldwide shipment of 10GBASE-T totaled a mere 182,000 units[21]. According to the Linley group, electromagnetic interference issues and large power consumption were cited as the main reasons for the poor uptake of 10GBASE-T devices.

As of now, no one is talking about 100Gigabit Ethernet over copper cabling. The exponential Ethernet speed growth over copper cabling is well and truly over. Nature sets physical limits and we have no option but to abide by them.

Therefore, it is time to stop expecting technological miracles to prolong the exponential growth phase of the human enterprise.

## 1.3 The Story of Inequality

It is that quest for growth that created Smalley's top ten list of humanity's "problems". These problems are systemic since their root causes are structural or cultural and therefore, largely independent of the individual actors. For instance, in the Oscar-winning movie, *Spotlight*, the psychiatrist and ex-priest, Richard Sipe, traces the sexual abuse scandals that are rocking the Catholic Church to the Church's requirement of celibacy for priests and its policy not to ordain women priests[22]. At any given time, roughly 50% of the priests are violating their vow of celibacy and consequently, this creates an environment of secrecy in which the abuses flourish. The abuses happened due to these systemic causes, and not because Catholic priests are more deviant compared to priests of other religions or denominations.

In December 2011, I witnessed an archetypal event that illustrated the systemic nature of our global socioeconomic and environmental predicaments. I was watching a village woman milk her cow in the village of Karech, adjacent to the Kumbalgarh Wildlife Sanctuary in the Udaipur district of Rajasthan, India. I have been working in

Karech since 2008 and I have received a more intense educational experience at Karech than all my years at Stanford!

Due to human activities including livestock grazing and firewood gathering, the Sanctuary, located on the border of the Thar Desert, is experiencing rapid degradation. People in this region of India are first hand witnesses to the major environmental catastrophes that the world faces today and their experiences are tremendously valuable for our understanding of these issues. Of the three major conventions that were adopted by the UN at the environmental summit in Rio de Janeiro in 1992, the UN Convention on Biological Diversity, the UN Convention to Combat Desertification and the UN Framework Convention on Climate Change, the villagers in Karech are experiencing all three environmental catastrophes on a daily basis[23]. Until about a century ago, these people were leading hunter-gatherer lifestyles, but they are now trying to adapt to a farming and herding lifestyle as the forest degrades around them. These days, families in Karech supplement their meager income by raising cattle for milk to trade with families in the neighboring town of Gogunda and by raising goats and selling them for export to the Middle East.

The woman began by untying the calf and allowing the calf to suck on the mother cow's udder. But within 30 seconds, she started pulling the calf away from the mother. The calf resisted her stoutly. The woman wasn't strong enough to pull the calf away and so she called her husband over. Between the two of them, they pulled the calf away from the udder and tied him in front of the mother. The calf was now bleating, obviously in distress, and the mother cow began licking her child. The woman milked the cow completely until there was nothing coming out of every teat in the udder. She extracted about 10 liters (2.5 gallons) of milk, sufficient to fetch her little more than lunch money for her family. She then released the calf to suck again on the udder and send a message to his mother's body that she doesn't have enough milk for her baby and needed to produce more.

All four sets of actors were suffering immensely in this drama:

1) the affluent consumer in the town of Gogunda who was suffering from obesity, diabetes and heart disease after consuming milk products;
2) the village woman who was desperately eking out a living;
3) the cow and her calf who were being ruthlessly exploited; and
4) the wildlife in the forest which was being starved to death, as the forest is constantly depleted when biomass and nutrients are eaten by livestock and shipped out to far away places in the form of milk and other livestock products.

While affluent consumers enjoy the material benefits of industrialization, they suffer from chronic diseases as well as social isolation and mental depression. Almost half the people in the US, the wealthiest country in the world, consume anti-depressants or anti-anxiety medications or mood-altering illegal drugs on a regular basis[24]. Wall street executives suffer one of the highest per capita rates of illegal drug use[25]. These consumers are largely disconnected from the direct consequences of their consumption as the deforestation and desertification happens out of their sight.

In contrast, village women in India enjoy the social benefits of a well-knit community, but they suffer from the environmental burdens of industrialization. The forests are dying, the desert is expanding, the temperatures are soaring, and the monsoons are erratic, mainly to meet affluent consumer demands. To top that, the firewood that villagers use for cooking produces smoke which now contain toxic industrial pollutants!

The toxic pollutants that we pump into the atmosphere in our industrial societies through burning fossil fuels and through our chemical processes, eventually come down in the rain, get absorbed by vegetation which filter these chemicals and store them in their trunks, branches and stalks. The village women are burning these tree branches and breathing the pollutants, while also recirculating the pollutants into the environment. As forests die out, there are fewer trees to do the pollution filtering and storage, which means that the concentration of these pollutants increases over time. Every year, we pour fresh toxic pollutants into the atmosphere.

Farm animals eat the vegetation and accumulate these toxic pollutants in their fat tissues as they grow. When affluent people consume livestock products, they ingest a concentrated dose of these pollutants, which they accumulate and store in their fat tissues as well. Thus, animal foods have become a source of numerous chronic diseases in affluent communities, since they effectively recirculate our industrial pollutants back to us. The USDA estimates that 95% of the dioxins in our bodies, which are some of the strongest carcinogens known to man, come from the foods we eat. Dioxins are released into the atmosphere whenever chlorine reacts with hydrocarbons and this happens, for instance, when we bleach wood pulp as consumers have been conditioned to prefer white paper over brown. The four main food sources of these dioxins are fish, eggs, cheese and meat, in that order[26].

There is no escaping the consequences of our actions, our Karma!

Thus far we relied on structural and cultural inequality both within and across species boundaries to fuel consumption growth leading to all that suffering. We have a financial system that largely originates new currency into the hands of the wealthy in the cities and in the global North and then trickles it down into the villages and the global South through economic transactions. We have inherited a culture that hierarchically layers people above animals and therefore pays scant regard to the well being of animals. As Dr. Paul Farmer said,

> "The idea that some lives matter less is the root of all that is wrong with the world."

While such structural inequality was perfect for fostering the exponential growth of the socioeconomic system, it is destroying the planet and failing us at the moment.

**1.4 The Story of Consumption**

Imagine going to a doctor with a persistent mild, $1^oC$ fever...

And a coconut-sized growth by the side of your head.

As you are waiting for the doctor to examine you, you sense a lot of nervousness in the office. You overhear the nurse furtively whispering to the doctor, "Don't mention the "C" word!" Then the doctor examines you and diagnoses that the growth is the cause of the fever. And that the fever is going to get worse! The best he can do is to limit your fever to $2^{O}C$ and maintain your health precariously in that advanced state of disrepair.

Would you then plead with the doctor to try and limit your fever to $1.5^{O}C$ and maintain it at that level?

Or would you ask him about "the C word," that ominous, coconut-sized growth by the side of your head, which he told you was the cause of the fever?

But when you ask him about the growth, imagine the doctor replies,

> "I will make sure that the growth doubles in size as quickly as possible. You will soon look like you have three heads! But your fever will be limited to less than $2^{O}C$".

Wouldn't you run away from such a doctor to seek a second opinion?

Now imagine that you have the same experience with the second doctor!

And a third!

You now feel as if the whole medical profession has gone berserk!

You then dig through the medical literature to understand the doctor's diagnosis about your condition. To discover that the growth can be reversed if you make some significant lifestyle changes. But the doctors didn't tell you that, because they were afraid that they would get their heads chopped off!

You see, it is impossible to get accurate medical advice when you are the Red Queen of Hearts in Alice's Wonderland!

Then you wake up.

And discover to your horror that the exact same scenario is being played out in the global environmental arena! Since the Rio summit, the UN Convention on Biological Diversity and the UN Convention to Combat Desertification have been abysmal failures, mainly because biodiversity loss and desertification don't affect the global North much. Most of the desertification and biodiversity loss is now happening in the global South, far away from the affluent consumers who are mainly responsible for it. Though climate change does affect the global North, even the UN Framework Convention on Climate Change has become an exercise in public relations as opposed to action. At the UN climate change conference, the twenty-first Conference of the Parties (COP-21), which concluded in December 2015 in Paris, the official story portrayed the Paris accord as a resounding success[27]. The nations of the world had voluntarily agreed to limit global warming increase to 2°C! As a stretch goal, and at the behest of the island nations of the world who fear being drowned in the rising seas, the nations of the world also agreed to do their best to limit global warming increase to 1.5°C!

Applause, applause!

But of course, it was just Kabuki theater. The political leaders of the world see no option but to keep the current fossil fuel drenched socioeconomic system chugging along until it collapses on its own. As Doug Carmichael puts it[28],

> "It is not irrational to stay in a leaky canoe if there is no other... It is better to keep going for another fifteen or twenty years and then collapse than to try to change it now and collapse now."

Indeed, the COP-21 negotiators had tacitly agreed not to mention the "C" word: Consumption. Four items were specifically off the table during the negotiations: animal agriculture, biofuels, aviation and shipping. In every future greenhouse gas emissions scenario that they considered, the negotiators also aimed to double the size of the global economy by the year 2100. Even though the global economy is currently estimated to be 60% larger than what the planet can support[29]!

Consumption was a taboo topic at the UN conference, specifically, the consumption of animal products. In fact, it was difficult to get good plant-based foods at the COP-21 venue in Paris, even though UN reports have repeatedly highlighted the adverse climate impacts of animal-based foods[30]. The negotiators themselves were a privileged bunch. There were hardly any women among them from all the 195 countries and by an insider's account, there was just one vegetarian among the whole lot.

The British climate scientist, Dr. Kevin Anderson of Cambridge University, noted the implicit savagery underlying the Paris accord, as he envisioned rich people in the global North muddling through and coping with the climate crisis as it unfolds, while the poor people in the global South were expected to die off[31]. He opined that nations in the global South could not possibly muster the resources to build sea walls and other strategies to improve the resilience of their societies in the face of rapid climate change, while the global North could afford to do so.

Personally, I think that Dr. Anderson has it backwards. He's assuming that the socioeconomic system will remain stable even as mass die-offs occur in the human population. Besides, it isn't just sea level rise that the nations of the world have to guard against with respect to climate change, but widespread disruptions in weather, seasons and rainfall patterns. In fact, the poor are resilient, live in strong communities and know how to grow their own food and can largely muddle through if climate change continues unchecked and the socioeconomic system collapses. It is the rich who are more vulnerable since they are isolated and dependent on external, corporate sources for their food, fuel and pharmaceutical intake. In exchange, the rich mainly possess pieces of paper with pictures of dead luminaries on them. Corporations and capitalism

in industrial societies as organized today would be difficult to
maintain in such unpredictable environments.

The eminent climate scientist, Dr. Jim Hansen, made headlines at
COP-21 when he said the idea that the world is making good
progress on climate change is "baloney"[32]. In his view, it is
possible to solve climate change but the nations of the world were
not advocating a solution in the Paris accord. Instead, he
recommended adopting a carbon "fee and dividend" scheme as the
more appropriate response to climate change. In this scheme, a
progressively increasing fee is collected at the source for all
carbon-based fuels and the entire fee is distributed as dividend to
all citizens. He estimated that the proposed carbon fee and dividend
approach would grow the world economy even faster than the
doubling that the Paris accord would accomplish by 2100!

Yes, dear readers, the cancer would grow even faster!

But we, the public, are largely playing the role of the Red Queen of
Hearts in Alice's Wonderland. We are complicit in this charade. No
political leader of ours would dare to reduce economic growth
without being summarily dismissed, anywhere in the world:

>    "Off with his head!"

No corporate CEO would dare to reduce profit growth without
being summarily dismissed by the shareholders and the board:

>    "Off with his head!"

No hedge fund manager would dare to reduce returns without being
sacked by his billionaire investors.

>    "Off with his head!"

No climate scientist working for the Intergovernmental Panel on
Climate Change (IPCC) would dare to suggest that solving climate
change might require reducing economic growth.

"Off with his head!"

It is fear that drives the current system, specifically the fear of the loss of the known, fear of the loss of stability of the socioeconomic system that we are ensconced in. Everyone is understandably paralyzed by the fear that they might topple the "leaky canoe" that they are precariously perched on. Economic growth is so central to our current way of organizing society that we are desperately trying to preserve it.

Growth is systemic. We have coded the quest for growth in the DNA of our currency structures so that the world's financial system will collapse without economic growth, or at least, the illusion of economic growth.

To foster that economic growth, we promote unnecessary, mindless consumption. Through advertising that pays for our "free" Internet, our "free" television programming, our "free" social media and our "free" newspapers. Consumption is the organizing value of our socioeconomic system as the average American is bombarded with 3500 advertisements each day.

Consumption is systemic.

To promote consumption, we encourage separation and isolation. Through glorifying the lifestyle choices of the loneliest groups of people, billionaires and celebrities, who are supposedly selected through fair market competition, the organizing principle of our socioeconomic system.

Separation is systemic.

At the heart of it is the story of separation from Creation, our Spiritual crisis. This is at the root of our suffering.

### 1.5 The Story of Separation

This story of separation is the core story that is truly failing us. With the technological strides we have made in the last two centuries, most of us live in concrete jungles with little to no exposure to the terrestrial biodiversity on Earth. Other than our pets, we rarely meet any other animal species in our daily lives except in zoos and circuses or packaged as meat in supermarkets. Therefore, many people have concluded that we are separated from Creation in an unconscious enactment of the Adam and Eve's Knowledge tree story from the Old Testament in the Bible, with the banishing of humans from the Garden of Eden. Said Charles Eisenstein, author of *The Ascent of Humanity* and *Sacred Economics*[33],

> "In civilization, what you are is a discrete, separate individual, among other individuals, in an external universe that is separate from you. In religion, you are a soul encased in flesh. In psychology, you are a mind encased in flesh. In biology, you are the expression of DNA serving to maximize your reproductive self-interest and greed. And that conception of self has basically poisoned our planet, because we treat it as if it were an other."

That is, not only are we separated from Creation, but we are separated from each other among our own species as well. Even the Interfaith Declaration on Climate Change signed by numerous faith dignitaries including Archbishop Desmond Tutu, reads in part[34],

> "While Climate Change is a symptom, the fever that our Earth has contracted, the underlying disease is the disconnection from Creation that plagues human societies throughout the Earth."

While this story of our separation, our atomization as a species, justifies and drives many of our daily actions, it is in fact a story of human exceptionalism, the idea that we are somehow different from and better than other species. It is based on the false notion that while other species all have to live in harmony with Nature, we are somehow exempt from that requirement since we can fashion our own environment.

This notion is patently false. The cascading environmental crises are signals from Nature that there are no such exceptions in the family of Life. We have no choice but to live in harmony with Nature because we are a part of Nature.

This story of separation is closely aligned with *speciesism*, which is discrimination and exploitation on the basis of species identity. It is due to speciesism that we consider the murder of humans to be wrong, but the hunting of other animals to be sport, concentration camps to be evil, but slaughterhouses to be humane, jails to be avoided, but zoos to be toured. This treatment of animals is at the heart of all our ecological crises, though most scientists studiously refuse to examine the connection for fear of treading on "values".

But fortunately, as will be shown in the next chapter, this story of separation is based on a delusion!

## 1.6 The Four Storylines

In addition to the stories we tell ourselves about the present, the stories that we tell to explain our past and the stories that we tell about our future also determine how we act in the present, which, ultimately, is all that matters. In general, there are four main storylines that we have been using to tell these stories in the context of the Earth and the environment. The first storyline is that,

**Everything is a mess and everything must change.**

This is the storyline that is most common in mainstream environmental circles today. As such stories go, our species made a huge mistake either 200,000-400,000 years ago with the discovery of the controlled use of fire as we evolved into our present form, or 10,000 years ago with the development of agriculture and city dwelling, or 200 years ago with the start of the industrial revolution[35]. The impact of that particular huge mistake is still reverberating around the planet and therefore, everything must change immediately. The first thing that we need to do now is to reduce human population from the present 7.4 billion to some number between 100 million and 2 billion, depending on the storyteller[36]. A prominent environmentalist even told me recently that humans would not make necessary lifestyle changes until the world human population plummets to 1 billion! At the recent COP-19 UN climate change conference in Warsaw, Poland, the Filipina Climate Chief, Mary Ann Lucille Sering, openly blurted out[37],

> "It feels like we are negotiating on who is to live and who is to die."

This promotes the idea that we are all waiting for the apocalypse to occur and take its toll before we become willing to change ourselves.

One of the major problems with this storyline is that it paints the majority of human beings, past and present, as mess-making creatures, which does not also inspire people to change. Consequently, much needed action still has not been taken, except for the inevitable jockeying for position in human societies. The rise of socialist, nationalist and xenophobic politicians in the global North is likely fueled by the suspicion that the ruling elites have formed secret alliances with ruling elites from other nations, while throwing the rest of their nation's citizenry under the bus.

The second, diametrically opposite storyline is that,

**Everything is perfect and nothing needs to change.**

The stories in this line also usually don't end well for most of our fellow beings on the planet either, except for those telling the stories and the believers in the particular supernatural deity that oversees the coming apocalypse. These true believers are the "winners" in these stories, while the other "losers" are swallowed up in a fire, or a flood, or a massive earthquake or some such calamity[38]. That's even more uninspiring than the first storyline!

The third storyline is that,

**Everything is a mess and nothing will change.**

Or rather, that while everything is a mess, we humans are too stuck in our habitual ways and nothing will change and that we might as well get used to the idea that we'll be going extinct. As the late comedian, social critic, actor and author, George Carlin, put it[39],

> "We are going away and we won't leave much of a trace either. Thank God for that! Maybe, a little styrofoam! Maybe, a little styrofoam! The planet will be here and we'll be long gone, just another failed mutation, just another closed end biological mistake, an evolutionary cul-de-sac. The planet will shake us off like a bad case of fleas, a surface nuisance."

While George Carlin predicted that all human beings will disappear, there are other story tellers who believe that this will result in a massive culling of the human population in a worldwide, dog-eat-dog type, apocalyptic cleansing following which life on Earth would recover, the human population would once again rebound to use up the recovering Earth's resources and the cycle would repeat. There are even scientific models predicting how such a roller coaster, life-and-death ride would unfold for us and our descendants over the next few centuries[40].

This storyline assumes that human beings are fundamentally no smarter than cyanobacteria in a petri dish, consuming, reproducing and perishing mindlessly. It is just as uninspiring as the previous two storylines.

The fourth storyline that is rarely, if ever, told, is that,

**Everything is perfect and everything will change.**

Or rather that everything is as it should be and as a result, everything will change. This book tells a story along this line for our species. The assumption of perfection helps guide this story towards the interpretation that despite the destruction that we've been doing to the planet's ecosystems, we do belong exactly as we are. It's just that we haven't yet understood the purpose of why we're doing what we're doing and this book advances one such plausible, evolutionary purpose.

I firmly believe that it is only such a positive framing of our past, present and future that can inspire the revolutionary changes called for today.

# 2. Separation is a Delusion!

*"We are what we think, all that we are arises with our thoughts, and with our thoughts, we make the world "* - Buddha.

In Dec. 2010, as I held our tiny, one-month old granddaughter, Kimaya, in my arms, surrounded by our family, two thoughts crossed my mind:

1. This was the most amazing thing that had ever happened to me, and
2. This baby girl must not inherit a trashed planet!

Three years earlier, I had been wallowing in the depths of despair, feeling an abject failure as a parent and as a despised member of a planet-destroying generation. But now, I saw Kimaya's birth as a sign of redemption, not just for our family, but for all humanity! Then a third thought occurred:

**What if everything is already perfect? Just as she was.**

I don't mean that in the sense of the mystics, as an article of faith. But rather as a statement of fact, based on science, reason and common sense[1].

What if we have been telling our human story all wrong?

We live on the most beautiful, life-sustaining planet that we could have ever imagined[2]. We are a truly privileged species within this amazing community called Life on Earth.

Therefore, despite the numerous difficulties that we face on the planet today, what if the world doesn't need to be changed? The world just needs to be understood. When we understand the world correctly, we will change ourselves, together. As we change, the world will change with us so that our difficulties will melt away.

The English writer and philosopher, Bertrand Russell, once said[3],

"The greatest challenge to any thinker is to state the problem in a way that will allow a solution."

But this can be interpreted in a reductionist sense. Unfortunately, that's how most of us have been educated to see the world, as full of smaller problems that need to be "solved". Bertrand Russell's dictum can even be stated as an algorithm, a sequence of actionable steps:

*Step 1*. List problems in order of importance.
*Step 2*. Solve them one by one.
*Step 3*. Reap the unintended consequences...
*Step 4*. Go to *Step 1*...

Is energy the number one problem for humanity? Then solve how can we produce 45% more energy by 2030. Next, solve how can we produce 30% more fresh water by 2030. And so on, down Smalley's list. Such reductionist thinking has led us to where we are today, with all the undeniable suffering that surrounds us. Therefore, what if the greatest challenge is to state our problems in a way that they are not problems, but indicators to personal and social change? As we respond to those indicators correctly, these problems will transform as well.

This is how flourishing ecosystems actually work. There is fluidity and flexibility on both sides spanning difficulties. After all, most of our problems are self-inflicted. Along with the problems, we have created enormous surpluses of profligate waste that we can harness to mitigate these problems as we change ourselves.

As William Ophuls wrote[4],

"The real product of genuine systems analysis is not solutions, but wisdom."

We have plenty of "solutions" for our "problems". Wisdom is precisely what we have been lacking, not solutions.

True wisdom leads to personal change, first and foremost.

## 2.1 The Perfection of Nature

It was the summer of 2009, a dark night in a remote wildlife sanctuary nestled in the Western Ghats of India. My sister, Sudha and I were walking back after dinner from the main house in the sanctuary to our detached sleeping quarters just a few hundred yards away. The sanctuary is not connected to the electric grid and therefore, we were carrying a solar torchlight between us to illuminate our way in the dark. That made it a very exciting short trek. Sudha is prone to be jumpy when insects land on her and the torchlight was attracting a whole bunch of them. When we reached our sleeping quarters, she bolted indoors, only to discover various insects had somehow managed to infiltrate the inside despite the mosquito netting on all the windows. Consequently, I had to run an informal taxi service at various times that night transporting insects from indoors to outdoors, which left me plenty of opportunities for reflection!

Our hosts, a couple from New Jersey, Pamela and Anil Malhotra, had been telling us the story of their Save Animals Intiative (SAI) Sanctuary over dinner[5]. They started the SAI Sanctuary by purchasing a 55-acre coffee plantation in the Kodagu district of Karnataka in 1991, tearing down the fences and just letting it exist in its natural state. From 1991 onwards, they had been steadily acquiring neighboring coffee plantations and tearing down fences so that wild animals could take refuge in their now 300-acre plus sanctuary. Sure enough, the animals did take refuge and those coffee plantations turned into the lushest tropical forest that you could imagine, all in the span of less than 20 years.

This was Eden, sheer perfection!

As Pamela described it, the forest regeneration was mainly due to the animals that wandered into the SAI sanctuary from the nearby National forests and then chose to make the sanctuary their home base. The elephants would eat ripe jackfruits in the National forests and then come back and deposit the seeds with their droppings in the sanctuary. The jackfruit tree is one of the primary tree species in the Western Ghat forests of India and its ripe fruit is a favorite snack for the elephant. Also, birds and other animals dropped other seeds over time and these diverse seeds combined with the rich soil and the monsoon rains to initiate the afforestation process. All Pam and Anil had to do was to patrol their land to keep the human poachers out and the animals safe. They didn't have to do much tree planting, tilling or fertilizing in order for the forest to regenerate. The elephants and other wildlife did most of it!

One of the advantages that the SAI Sanctuary enjoyed was its close proximity to the Bandipur and Nagarhole National forests and the Mudumalai wildlife sanctuary. However, the biodiversity found in the SAI sanctuary is far richer than the biodiversity in these nationally protected forests mainly because these protected forests are also "managed" by the Forest Service of India. As any visitor to these nationally protected lands would attest, this "management" has resulted in rows and rows of neatly planted teak and eucalyptus trees, which are useful for furniture and lumber and count as tree cover in forest land surveys, but are otherwise out of place in the forests of the Western Ghats.

At the SAI sanctuary, the trees were growing randomly, not in neat rows. The underbrush was so thick that only creepy, crawly creatures could traverse through them. The main paths in the forest were the ones made by the elephants who are the only wild animals capable of clearing such paths in the forest. All the other animals including us, humans, followed these elephant paths. Elephants also casually break branches from trees to eat the leaves while creating openings for sunlight to stream down and nourish the underbrush. Everything that the elephants did in the forest seemed to have some beneficial impact on the forest. It was as if the elephants knew exactly what to do to be an asset to the forest ecosystem, to keep life flourishing. They belonged! So did all the other animals. In contrast, it seemed that the best that we, human beings, could do to be an asset to the forest ecosystem, to keep life flourishing, was to simply stay away from the forest.

This troubled me as I was running that informal insect taxi service for my sister's benefit. I was also born in the same forest, some 200 miles away from the SAI sanctuary and some fifty years before, but why don't I belong as is in the forest ecosystem? How did I grow up to be such an outcast in my own home, my birthplace?

This is a common theme in many environmental circles, that human beings, with those in traditional indigenous cultures exempted, are a destructive force in the Earth's many ecosystems. We don't belong. In contrast, every creature in that sanctuary seemed to have found an ecological niche so that the night sounds of the forest filled every octave of my auditory spectrum. Clearly, each had a distinct identity within that ecosystem. Between them, these creatures had turned a coffee plantation into such a thriving forest that students of ecology from reputed American universities were spending months at the SAI sanctuary studying the biodiversity of the Western Ghats. The wildlife routinely contributed more to the recovering ecosystem than they consumed from it until the ecosystem flourished and reached a thriving stable state. In contrast, we humans tend to consume far more from even a flourishing ecosystem than we contribute to it and thereby cause its rapid degradation.

So wild animals live their normal lives and the forest thrives. Humans, especially those of us in the cities of the world, live what we consider to be our normal lives, and not just the forest, but the entire planet dies?

That's such a depressing story to live by. Surely, we should be telling better stories about ourselves, about whom we are and what we need to do going forward, based on scientifically verifiable facts? Surely, there is an identity, an ecological niche that we can assume as a species, so that we belong on the planet exactly as we are? Sustainability is attained when we can routinely contribute more to ecosystems than we consume from them, just as the elephants do for the forests of the Western Ghats and indigenous communities do for the Amazon rainforest. Surely we can devise systems of social organization that can help us achieve that in our technological civilization? What common story should we live by that can help us reach that goal?

Such a common story was staring me in the face at SAI Sanctuary though it took me five long years to piece it together. Pam and Anil were shining examples of how we can live in harmony with Nature. Compassion for all Creation was the core organizing value in their simple lives, not mindless consumption. In contrast, the poachers who had to be kept out of the SAI Sanctuary were desperately eking out a living by meeting the demand for ivory from remote consumers who had bought into that consumption paradigm.

It was the Karech milk story all over again!

If we're so inclined, we can read the rest of the book assuming that we are each part of a perfect whole and therefore, there's an infinitely compassionate, higher intelligence at work in the cosmos at large. Call this higher intelligence God, Yahweh, Allah, Brahman, the Creator, the Great Spirit, Higher Consciousness, or just Life, that amazingly beautiful process which seemingly defies the fundamental laws of entropy (it doesn't) and coaxes order out of chaos. It is the same holistic intelligence, which guides the elephant, the tiger, the birds and the insects to routinely create the sheer perfection of the SAI sanctuary. True, there is suffering in the SAI sanctuary as predators consume prey, but Life as a whole thrives in that beautiful intricacy of mutual connectedness. It is surely unimaginable that this higher intelligence would birth my fellow human beings and I in the perfection of those same forests of the Western Ghats - in order to destroy it?

That story of separation is rooted in exceptionalism! Instead, let us begin with the assumption that everything is indeed perfect as is. This is also in alignment with the undeniable material progress that we have already made during the course of history! As Peter Diamandis, the chair of Singularity University and co-author of *Abundance*[6], pointed out in his TED talk from 2012[7],

> "Over the last hundred years, the average human lifespan has more than doubled, average per capita income adjusted for inflation around the world has tripled. Childhood mortality has come down a factor of 10. Add to that the cost of food, electricity, transportation, communication have dropped 10 to

1,000-fold. Steve Pinker has showed us that, in fact, we're living during the most peaceful time ever in human history. And Charles Kenny that global literacy has gone from 25 percent to over 80 percent in the last 130 years. We truly are living in an extraordinary time. And many people forget this. And we keep setting our expectations higher and higher. In fact, we redefine what poverty means. Think of this, in America today, the majority of people under the poverty line still have electricity, water, toilets, refrigerators, television, mobile phones, air conditioning and cars. The wealthiest robber barons of the last century, the emperors on this planet, could have never dreamed of such luxuries...

When I think about creating abundance, it's not about creating a life of luxury for everybody on this planet; it's about creating a life of possibility. It is about taking that which was scarce and making it abundant. You see, scarcity is contextual, and technology is a resource-liberating force...Think about it, that a Masai warrior on a cellphone in the middle of Kenya has better mobile comm than President Reagan did 25 years ago. And if they're on a smartphone on Google, they've got access to more knowledge and information than President Clinton did 15 years ago. They're living in a world of information and communication abundance that no one could have ever predicted. Better than that, the things that you and I spent tens and hundreds of thousands of dollars for -- GPS, HD video and still images, libraries of books and music, medical diagnostic technology -- are now literally dematerializing and demonetizing into your cellphone".

While some of this material abundance occurred for exploitative reasons, e.g., the cell phone is a tool to summon distant labor to work in the cities at slave wages, we cannot deny that it has occurred. Our communications infrastructure has turned this world into a neighborhood laying the framework for our ethical commitment to turn it into an Eden. Even on a moral and ethical basis, we have made tremendous strides in the industrial era. Take for instance, our granddaughter, Kimaya. She is half Indian, one-quarter African American and one quarter American Indian. Thus, she is the daughter of three continents, born in America.

A hundred and fifty years ago, she might have been captured and brought on a slave ship from Africa[8].

A hundred years ago, she might have been a colonial subject in India[9].

Seventy-five years ago, she might have been a zoo exhibit in Europe[10].

Fifty years ago, she might have been forcibly educated out of her cultural heritage in America[11].

Yet, she was a welcome guest at the European Parliament just prior to COP-21 in Dec. 2015[12]. This speaks volumes about the ethical and moral strides we have made in just a few generations.

## 2.2 The Caterpillar and the Butterfly

In my 2011 book, *Carbon Dharma: The Occupation of Butterflies*[13], I used the metaphor of the Caterpillar and the Butterfly to recast the human story in a positive light, where we do belong exactly as we are. It is a metaphor that has been used by numerous other authors before me. It is an apt metaphor based on the observation that we have currently organized our society around consumption as the core value and we are transitioning towards a society organized around compassion as the core value. As Judith Anodea writes in her book, *Waking the Global Heart* [14],

> "When a caterpillar nears its transformation time, it begins to eat ravenously, consuming everything in sight. The caterpillar body then becomes heavy, outgrowing its own skin many times, until it is too bloated to move. Attaching to a branch (upside down, where everything is turned on its head), it forms a chrysalis—an enclosing shell that limits the caterpillar's freedom for the duration of the transformation.

Within the chrysalis a miracle occurs. Tiny cells, that biologists call "imaginal cells," begin to appear. These cells are wholly different from caterpillar cells, carrying different information, vibrating to a different frequency—the frequency of the emerging butterfly. At first, the caterpillar's immune system perceives these new cells as enemies, and attacks them, much as new ideas in science, medicine, politics, and social behavior are viciously denounced by the powers now considered mainstream. But the imaginal cells are not deterred. They continue to appear, in even greater numbers, recognizing each other, bonding together, until the new cells are numerous enough to organize into clumps. When enough cells have formed to make structures along the new organizational lines, the caterpillar's immune system is overwhelmed. The caterpillar body then becomes a nutritious soup for the growth of the butterfly.

When the butterfly is ready to hatch, the chrysalis becomes transparent. The need for restriction has been outgrown. Yet the struggle toward freedom has an organic timing."

Imagine the Caterpillar that has engorged itself to bursting point. Entering the pupal stage, its world is turned upside down.

War is peace.

Debt is wealth.

Slavery is freedom.

Misery is happiness.

Isolation is friendship.

The Caterpillar struggles to maintain its growth phase as it encounters the physical limits of the planet. The dominant class throws up one last "savior," one who promises to make it great again, to restore its glory days when the hierarchical, authoritarian lines were clearly drawn[15]:

God above Man.

Man above Woman.

Whites above Coloreds.

Straights above Gays.

Winners above Losers.

Rich above Poor.

People above Animals.

Us above Them.

But this is the last hurrah of the Caterpillar as it tries to cope with the systemic symptoms of its engorged state.

Climate change is systemic.

Biodiversity loss is systemic.

Desertification is systemic.

Toxic pollution is systemic.

Inequality is systemic.

Poverty is systemic.

Slavery is systemic.

Sexism is systemic.

Racism is systemic.

Casteism is systemic.

Homophobia is systemic.

Speciesism is systemic.

Any socioeconomic system that is organized around consumption as a core value and competition as a core principle is bound to exhibit these symptoms. But the metamorphosis is inexorable. "Imaginal cells" addressing each of these symptoms have sprung up throughout its body politic. These imaginal cells are the precursors of the Butterfly. The Caterpillar's system has been in a continuous state of war against these cells. One by one, the battles for ideas have been lost by the Caterpillar.

Slavery is now politically incorrect.

Sexism is now politically incorrect.

Racism is now politically incorrect.

Casteism is now politically incorrect.

Homophobia is now politically incorrect.

Though these oppressions are still continuing in society, the Caterpillar now has to use guarded language and dog whistle politics to exploit them. On the surface, it promotes "equality", meaning a level playing field for the competitions that determine its privileged classes. But environmentalism and speciesism are still considered fair game for open repression in mainstream circles. This is where the Caterpillar is mounting a last-ditch defense of its divide and conquer strategy to promote its endless growth ideology. Worldwide, over 100 environmental activists have been murdered annually, culminating in the brutal assassination of Berta Caceras and her colleague, Nelson Garcia, in Honduras[16] in March 2016. In the US, the Caterpillar has openly designated environmental activists and animal rights activists as the number one domestic terrorism targets of the Federal Bureau of Investigation (FBI)[17]. Such activists now all have dossiers, their every move watched.

It is illegal to take photographs or videos of animal enterprise operations!

It is illegal to know where your food comes from!

It is illegal to picket oil pipelines!

It is illegal to quench the thirst of slaughterhouse bound animals!

It is illegal to do anything that would reduce the profits of corporations!

But the imaginal cells are becoming too numerous. They are overwhelming the Caterpillar's immune system. They are clumping together to form coherent groups, precursors of the Butterfly's organs. They identify the limits of the chrysalis: no more than half the land area of the Earth must be used for human purposes, while the other half must be returned to Nature to let biodiversity recover and flourish[18]. They recognize that in reality, it isn't the endless economic growth of the Caterpillar that we truly want, but economic security.

It isn't unlimited consumption that we truly want, but unlimited happiness.

It isn't the pangs of isolation that we truly want, but the bonds of community.

Thus from a systems perspective, the metamorphosis from the Caterpillar to the Butterfly requires much more than just restoring the balance in all the major bio-geophysical cycles of the planet, the carbon cycle, the nitrogen cycle, the phosphorous cycle, the hydrological cycle, the species birth-extinction cycle, the materials cycle, and so on. The Butterfly is a state of being that can persist forever and is therefore, infinitely sustainable. Hence it is a state of "moral singularity," in which liberty, equality and the pursuit of happiness are a lived reality for all of humanity, not just words in our national constitutions.

For freedom is infinitely sustainable. Slavery is not.

Equality is infinitely sustainable. Inequality is not.

Peace is infinitely sustainable. War is not.

Justice is infinitely sustainable. Injustice is not.

Happiness is infinitely sustainable. Misery is not.

Compassion for all Creation is infinitely sustainable. Violence towards any part of Creation is not.

In short, all that we know as "good" is infinitely sustainable. All that we know as "evil" is not.

Therefore, the metamorphosis calls us to do social engineering as well as engineering of the material kind. Social justice activists of all stripes, gender rights activists, civil rights activists, LGBT activists, environmental activists and animal rights activists, are all fellow seekers in our global quest for sustainability today. This moral aspect is why the major religions of the world have been weighing in on climate change, with prominent declarations from the Hindu[19], Christian[20], Islamic[21], Baha'i[22], Buddhist[23], Jewish[24], Sikh[25], Unitarian Universalist[26] and Interfaith[27] communities, exhorting all adherents to take action.

At first, the imaginal cells, the social justice activists, need to set up a framework to ensure that they operate within the limits of the chrysalis. They must unite together under the organizing value of compassion as opposed to consumption, so that they consciously cease to enslave and exploit animals. This will free up land, water and energy resources that are being currently wasted for Animal Agriculture to heal the planet during the metamorphosis. Then the Butterfly will be ready to break free.

Once the Butterfly is born, there will be no need for the limits of the chrysalis. For the Butterfly is fundamentally a life-sustaining presence. Then the Butterfly will wield all the tools and technologies that the Caterpillar had developed to fulfill its life-sustaining purpose.

This is the crucial moment of uniting the two processes: waste and transformation, almost like Yoga, literally, "union." As the ancient Yogis have repeatedly said, separation is a delusion, the greatest of all the delusions that the human ego has used to nourish its false, separate identity[28].

Nothing in Nature is ever separate from Nature. Human beings are no exception. We belong in Nature and we belong exactly as we are!

### 2.3. The Indian Ending[29]

The story of separation which has animated Western civilization and which has now been imposed globally, must now end in a common story of reunion and redemption for all humanity. Edward O. Wilson, the eminent ecologist and Professor Emeritus of Biology at Harvard University, has been a champion of biodiversity for decades, and he recently enunciated the Half-Earth strategy for the regeneration of Eden, which humanity must adopt in our chrysalis phase[30]. Such reunion and redemption is the vision for humanity that the great 20th century seer and philosopher, Sri Aurobindo, espoused in his epic poem, *Savitri*[31].

*Savitri* is based on a story told in the *Vana Parva* of the Hindu epic, *Mahabharata*[32]. Savitri is a woman who takes birth as the daughter of a spiritually disciplined father, the King of Madra, Aswapathi. Savitri is so beautiful, pure and ascetic that no man dares to come forward to ask her hand for marriage. Therefore, Savitri sets out to choose her own husband.

She finds and marries Satyavan, the son of the blind king Dyumatsena, who has lost not just his sight, but his kingdom as well and is now in exile. The sage, Narada, tells Savitri that her chosen husband will die on their first wedding anniversary. But Savitri is determined to save her husband's life. When Yama, the God of Death, arrives to take Satyavan's life in the forest, she pursues him and argues with him until he relents and blesses her with eternal happiness. Then Satyavan awakens as though from a deep sleep to find Savitri by his side. When they return home, they discover that Satyavan's father has his kingdom and sight restored as well.

As with all the stories in the Mahabharata, Savitri is a symbolic myth. Satyavan is the human spirit descended into ignorance and dissolution, born of the blind mind, Dyumatsena. Savitri is the Divine Word, born to save the human spirit. Aswapathi, her father, is the concentrated energy of spiritual endeavor that births the Divine Word. As Sri Aurobindo asserts in the introduction to his epic poem[33],

> "This is not a mere allegory, the characters are not personified qualities, but incarnations or emanations of living and conscious Forces with whom we can enter into concrete touch and they take human bodies in order to help man and show him the way from his mortal state to a divine consciousness and immortal life."

*Savitri* is pointing out that our feminine side has the power to redeem our fallen spirits. Indeed, women in the global North already have the power to transform the socioeconomic system since they control over 80% of all purchasing decisions[34]. *Savitri* is one of those rare myths that postulate Heaven can be right here on Earth. Who wouldn't want to work towards that?

As the noted British historian, Arnold Toynbee, is reported to have said[35],

> "It is already becoming clear that a chapter which had a Western beginning will have to have an Indian ending if it is not to end in self-destruction of the human race. At this supremely dangerous moment in human history, the only way of salvation is the ancient Hindu way. Here we have the attitude and spirit that can make it possible for the human race to grow together in to a single family".

The Western beginning of this industrial civilization chapter is characterized by scientific discovery and a relatively free and advanced outer material life, but a relatively rigid and doctrinaire inner spiritual life. The ancient Hindu way is characterized by spiritual discovery and a relatively free and advanced inner spiritual life, but a relatively rigid and caste-based outer material life. As we now combine the best of both worlds, we can usher in an era of material and spiritual prosperity for all of humanity and ensure the flourishing of all Life. This is precisely what we are called to do in this axiomatic summary of the Bhagavad Gita:

"Everything that has happened, has happened for the best. Everything that is happening, is happening for the best. Everything that will happen, will happen for the best."

For what Arnold Toynbee calls "the ancient Hindu way" is based on the simple idea that we are already living in a world of perfection.

# 3. Everything is Perfect!

*"Science without religion is lame. Religion without science is blind"* -
Albert Einstein.

In our Caterpillar phase, the human ego has risen high as we have
become more and more deluded that we are separated from Nature.
As a result, even our stories about supernatural deities have become
tinged with human hubris.

Do you believe in one God? If so, then regardless of your faith
tradition or denomination, it is likely that your God is omnipotent,
omniscient and omnipresent.

In the Christian Bible, Job acknowledged God's omnipotence in
Job 42:2[1],

> "I know that you can do all things and that no plan of yours can
> be thwarted."

In Islam, Allah is the Supreme Power, the Creator, the All Mighty
and the All Merciful[2].

In the Taittiriya Upanishad of the Hindu Yajur Veda, God is defined
as[3]

> *"Sathyam Jnanam Anantham Brahman* (Brahman is Reality and
> Knowledge Without Limits)."

In the remaining verses, the Upanishad then logically deduces that
Brahman is existence itself, consciousness itself and happiness
itself. Such a definition leads to universality. Indeed, since
Brahman is existence itself and consciousness itself, everyone,
including atheists, can accept that Brahman exists! After all,
atheism is the denial of the existence of a straw man version of God
imagined to be a kindly, white bearded man who lives in the sky.
Instead, Brahman is defined to be an immanent presence in our

daily lives. Further, since Brahman is happiness itself, everyone can accept that all beings seek Brahman at all times!

As a corollary, the Taittiriya Upanishad deduces that Brahman is eternal, omnipresent, omniscient and omnipotent.

Therefore, even though our religious differences have been used to divide us in the Caterpillar phase, the Christian God, the Islamic Allah and the Hindu Brahman are essentially one and the same. This means that for centuries, we have been arguing about nomenclature and not substance! But, if this common deity of ours is omnipotent, then the question arises as to how can such a deity be powerless in the face of so-called "human abuse" of the planet? Indeed, the Papal Encyclical, Laudato Si, contains the following admonishment[4]:

> "Our sister (The Earth) now cries out to us because of the harm we have inflicted on her by our irresponsible use and abuse of the goods with which God has endowed her. We have come to see ourselves as her lords and masters, entitled to plunder her at will. The violence present in our hearts, wounded by sin, is also reflected in the symptoms of sickness evident in the soil, in the water, in the air and in all forms of life."

In the Islamic Declaration on Climate Change as well, the tone is one of recrimination towards human behavior[5]:

> "Our species, though selected to be a caretaker or steward (*Khalifah*) on the earth, has been the cause of such corruption and devastation on it that we are in danger of ending life as we know it on our planet."

Finally, in the Hindu Declaration on Climate Change, the signatories warn humanity[6]:

> "Rapacious exploitation of the planet has caught up with us. A radical change in our relationship with nature is no longer an option. It is a matter of survival. We cannot destroy nature without destroying ourselves."

The undercurrent in these declarations is that humans have been plundering, corrupting and rapaciously exploiting the Earth, in direct opposition to God's will. That is, humans have become so powerful today that even an omnipotent deity is overwhelmed by our collective strength!

Such hubristic thinking is reflected in the oft-quoted reaction of the nuclear scientist, Dr. Robert Oppenheimer, who thought when he witnessed the world's first nuclear explosion[7],

"I am become Death, the Destroyer of Worlds!"

He was channeling the words that God spoke to Arjuna, the human protagonist, in the Bhagavad Gita, the Hindu Scripture.

Therefore, even scientists have been channeling this human hubris.

### 3.1 The Purpose of Life

Our egotistical thinking is then reflected in the attributes that we assign to our deities. The Biblical verse, Isaiah 43:7[8], states:

"Everyone who is called by My name, And whom I have created for My glory (my emphasis), Whom I have formed, even whom I have made."

This is commonly interpreted to mean that the purpose of human life is to sing God's praises. But is an omnipotent deity so lacking in self-worth that a distinct species had to be created on Earth for that special purpose? Or is our interpretation of that line in the Bible just a reflection of our own thinking patterns?

Science has not been of much help in this regard either. As the historian, Yuval Noah Harari, puts it[9],

"As far as we can tell, from a purely scientific viewpoint, human life has absolutely no meaning. Humans are the outcome of blind evolutionary processes that operate without goal or purpose."

But do evolutionary processes truly operate without goal or purpose? Then, what made the SAI Sanctuary flourish in such a short period of time?

Google's Artificial Intelligence "chatbot" has come the closest to making sense on this question[10]. Here is a relevant snippet of its responses to questions:

**Human:** "What is the purpose of life?"
**ChatBot:** "To serve the greater good."
**Human:** "What is the purpose of living?"
**ChatBot:** "To live forever."

The Google ChatBot's answers are clearly in alignment with Life's processes as we observe them in Nature. Life wants to live forever. Life has been on Earth continuously for 3.6 billion years, which is unimaginable for a random process without any intrinsic impetus to live. Indeed, every life form of significance fights death. You can witness it in the slaughterhouses where even baby animals fight for their lives. You can witness it in humans who go to great lengths to keep themselves alive[11]. You can witness it in the SAI Sanctuary where just leaving Nature alone was sufficient to cause the forest to bounce back.

Every viable species contributes to the greater good of all Life on Earth. Perhaps it is the glorification of all Creation that is called for in the Biblical verse quoted above and not just expressing words of praise to an insecure deity? We can truly "praise God" only when we help Nature thrive.

In the Hindu view, the purpose of Life is to transcend earthly bonds and to realize the Supreme Brahman with certainty in every fiber of your being. When you feel your existence in every pore of your body (Sat), when you feel one with the consciousness of the whole universe (Chit) and when you tingle with bliss in every waking moment (Ananda), then you are truly enlightened. But to become so enlightened, the Hindu is expected to traverse through the Earthly stage where he or she pursues, in order of importance,

1. *Dharma*, working for the greater good;
2. *Artha*, using whatever skill he or she possesses;
3. *Kama*, with pleasure and pain as a guide; and

Spiritual enlightenment, or "*Moksha*", is the fourth and final stage of this orderly four-step process[12]. As we shall see, it is this same sequence of steps that the human species has been traversing in order to reach for our collective state of moral singularity, our *Moksha*.

## 3.2 Compassion for all Creation

At their core, all faith and wisdom traditions, including secular humanism, contain the same nugget of truth that only compassion for all Creation helps Creation thrive, whereas unbridled violence to any part of Creation cannot endure. At the 2011 UN Climate Change Conference, the 17th Conference of the Parties (COP-17), forty prominent thought leaders from various faith and wisdom traditions, Christian, Muslim, Hindu, Buddhist, Jewish, Sikh and Bahai, including Archbishop Desmond Tutu, signed an Interfaith Declaration on Climate Change, which contained the following addendum[13]:

> "While climate change is a symptom, the fever that our Earth has contracted, the underlying disease is the disconnection from Creation that plagues human societies throughout the Earth.

> We, the undersigned, pledge to heal this disconnection by promoting and exemplifying compassion for all Creation in all our actions."

Thus compassion for all Creation is indeed compatible with every major faith and wisdom tradition in the world. His Holiness the Dalai Lama once told his 4 million Facebook friends[14],

"All the world's major religions with their emphasis on love, compassion, patience, tolerance and forgiveness can and do promote inner values. But the reality of the world today is that grounding ethics in religion is no longer adequate. This is why I am increasingly convinced that the time has come to find a way of thinking about spirituality and ethics beyond religion altogether."

He used the analogy of tea and water, with water being a secular ethic, say compassion, while tea is a religion that espouses it. He said,

"But however the tea is prepared, the primary ingredient is always water. While we can live without tea, we can't live without water. Likewise, we are born free of religion, but we are not born free of the need for compassion."

We need compassion as much as we need air. That is also at the core of our quest for sustainability:

**The Law of Sustainability**

**Compassion for all Creation is infinitely sustainable. Conversely, violence to any part of Creation is unsustainable.**

Sustainability means that the particular activity can be continued indefinitely and without limits. Sustainability of a human presence on Earth is achieved when we routinely contribute more to the Earth than we consume from the Earth, like the elephants at SAI sanctuary. Compassion for all Creation or kindness to all Life is summarized in a single, ancient Sanskrit word, "Ahimsa"[15] and clearly, we can manifest this without limit and sustain it. Conversely, violence to any part of Creation, or "Himsa" in Sanskrit, is unsustainable, meaning it will stop, either when that part of Creation is destroyed or we go extinct in the process of committing that violence persistently. Usually, the persistent nature of the violence manifests itself as the "Tragedy of the Commons" and need not be espoused explicitly. If the violence is instituted in a capitalist system, then the quest for infinite profit would take us to the brink. If the violence is instituted in an autocratic system, then the quest for infinite power would take us to the brink. Finally, if the violence is instituted in a socialist system, then Jevon's paradox would take us to the brink there as well.

If every human being routinely lets compassion for all Creation guide all actions as the Durban signatories pledged, then our world would be idyllic indeed.

### 3.3 The Paradox of Perfection

I contend that in an inexorable march towards that idyllic world, we had to create a mess to begin with! This is an extraordinary claim which requires extraordinary factual evidence.

In his book, *Awareness*, the great Jesuit priest and mystic, Fr. Anthony DeMello, wrote[16],

> "You know, all mystics—Catholic, Christian, non-Christian, no matter what their theology, no matter what their religion—are unanimous on one thing: that all is well, all is well. Though everything is a mess, all is well."

Such mystics have experienced the truth and have therefore, cleansed themselves of their subliminal fears and self-recriminations. When all is well within, everything is perfect without.

But is there a scientific basis for this assertion of perfection in the face of all evidence to the contrary? Can we indeed show that our delusion of separation from Nature was a necessary step in human evolution, as part of the compassionate perfection of reality?

Science searches for the truth through patterns in common experience. But nothing is sacred in science and skepticism is fundamental to progress in science. The eminent physicist and Nobel Laureate, Richard Feynman, said in an address to science teachers[17],

"Science.. contains within itself the lesson of the danger of belief in the infallibility of the greatest teachers of the preceding generation."

Thus, questioning everything is the bedrock principle of science. Can our assertion of perfection withstand the cold, hard test of such skeptical, scientific scrutiny? Can we satisfactorily and logically explain all the seeming ills of our present human condition as having occurred for a common higher, teleological purpose? If so, for what purpose?

Can we see perfection in colonialism? Why did my South Asian Indian ancestors have to endure the indignity of colonialism in such a world of perfection? Indeed, colonialism is still prevalent today, though couched in more polite euphemisms. The indigenous peoples of the world are as colonized today as my ancestors were in the heyday of the British Raj. At present, indigenous people have little to no say in their own homelands, their treaties are routinely ignored, their homes destroyed, their wives and children raped and the rivers and lakes in their homelands are polluted without their consent, all in the name of progress[18]. The industrial world can't convert millions of acres of forests into grazing lands for cattle, soy fields for livestock feed in the Amazon, palm oil plantations in Indonesia and new mining strips in the Congo, every year, without trampling on the rights of indigenous people violently[19]. Even the newly independent states in the post-colonial era have been ensnared with debts through development projects under the Breton-Woods arrangements and made to cough up their natural resources as reparation[20]. That is economic colonialism, even if the ex-colonial masters no longer rule these nations.

Can we see perfection in slavery? Though human slavery is no longer legal in most parts of the world, it still persists in the illicit sex industry, in the recidivist prison population of the US, in bonded, child labor and in the forced labor pools of China. While African Americans constitute just 13% of the US population, they constitute almost half the prison population of the US[21]. According to Steve Fraser and Prof. Joshua Freeman, in the US[22],

"Nearly a million prisoners are now making office furniture, working in call centers, fabricating body armor, taking hotel reservations, working in slaughterhouses, or manufacturing textiles, shoes, and clothing, while getting paid somewhere between 93 cents and $4.73 per day,"

That is thinly disguised slavery! Besides, consider that all the people of the world subjected to total Internet surveillance by large corporations and the National Security Agency (NSA) of the US, lack free agency. Says Prof. Eben Moglen of Columbia University[23],

"We've lost the ability to read anonymously. Without anonymity in reading there is no freedom of mind, there's literally slavery."

Since Yahoo, Google, Facebook, Microsoft and the National Security Agency (NSA) are constantly recording and watching our every move online, they have literally enslaved every Internet user on the planet. Such ubiquitous surveillance is enslavement because there are thousands of obscure national, regional and international laws on the books that make each and every one of us vulnerable to prosecution for seemingly innocuous acts. For instance, in the US, the Lacey act, 16 U.S.C #3371-8 states[24]:

"It is unlawful for any person... to import, export, transport, sell, receive, acquire or purchase any fish or wildlife or plant taken, possessed, transported, or sold in violation of any law, treaty, or regulation of the United States or in violation of any Indian tribal law or in violation of any State law or in violation of any foreign law."

Suddenly, even our email exchanges over last night's dinner could be grounds for our prosecution and conviction.

Can we see perfection in racism? Despite the passage of the Civil Rights Act of 1964 in the US, which made discrimination on the basis of race illegal, institutional racism still persists in the common day to day interactions between people of different races.

The police routinely arrest more African Americans and American Indians for the same crime than White Americans, despite studies, which show that statistically, these groups commit crimes at roughly the same rate. African Americans are 4.4 times more likely to be arrested for property offenses, 6.4 times more likely to be arrested for violent offenses and 9.4 times more likely to be arrested for drug offenses[25]. Such racism is flourishing not just in the US, but also around the world. Neo Nazi groups are openly active in many countries, promoting hate on the basis of skin color and other such superficial externalities[26].

Can we see perfection in casteism? Casteism, or discrimination on the basis of birth lineage, is the uniquely Indian version of racism. Though casteism is officially frowned upon in India, tradition has kept the fire of casteism burning even as it consumes the intellectual treasures of India, the creativity of her people[27].

Can we see perfection in ableism, in society's discrimination on the basis of ability? Can we see perfection in ageism, in society's discrimination on the basis of age? Even though most countries have outlawed such discrimination, they are still common in our daily lives[28].

Can we see perfection in sexism, in society's institutionalized, paternalistic discrimination against women? Though sexism is largely condemned in society's legal strictures, it is still an all too common occurrence. Says Lori Girshick[29],

"When people watch a video of abused women or hear abused women and men speak about the beatings, rapes, and dominations experienced at the hands of partners and ex-partners, these people are clearly moved with compassion for their plight. However, feeling compassion for these individual survivors is very different from understanding that there is a social system that influences and condones this violence, a legal system that inadequately addresses it, a media system that encourages power-over, and a sexist belief system inherent in the religious teachings, gender roles, and traditions that form the context in which we all operate."

Can we see perfection in poverty, hunger and the inequity that plagues human societies today? While nearly a billion people go hungry around the world with 45% of the children below the age of 5 malnourished in India, more than a billion people around the world are overweight or obese and clearly, don't want to be either[30]. While Mukesh Ambani, the Indian billionaire, has built a $1B 27-story home in the middle of Mumbai, India, for himself and his four family members, millions of people stew in slums just a few miles from his opulent home[31].

Can we see perfection in homophobia, one of the last, legally sanctioned discriminatory practices against other human beings within human societies[32]? The struggles of our Lesbian, Gay, Bisexual, Transsexual and Queer (LGBTQ) brethren to achieve legal equality are still ongoing in many conservative societies, though progress is being made at a rapid pace as of late. But as our experience with racism and sexism shows, legal equality is just the beginning of a long continuous struggle for lived equality.

Can we see perfection in speciesism, literally the mother of all the violent, discriminatory practices in human societies? While many people would fundamentally agree with the concept of equality for all humans, they would still mentally classify animals as different; to be used as we please. But as we have progressively enlightened ourselves over the years to accord equal consideration to all humans regardless of race, color, creed, caste, sexual orientation or gender identity, isn't it time that we enlighten ourselves to accord equal consideration to all sentient beings? Don't all sentient beings deserve to live free of socially sanctioned, deliberately inflicted, misery and suffering? The Australian philanthropist, Philip Wollen, has asserted that[33],

"Animal rights is now the greatest social justice issue since the abolition of slavery."

Speciesism is the belief in the superiority of one species, namely human beings, over all other species on the planet. It is a pervasive, legally sanctioned violence, perpetrated routinely by billions of people all over the world, without a second thought, at least three times a day. In most societies, eating animal foods is considered "Normal, Natural or Necessary," Dr. Melanie Joy's three N's of justification, in any system of oppression[34]. As the writer, philosopher and holocaust survivor, Isaac Bashevis Singer, wrote in his book, Enemies: A Love Story[35],

> "As often as Herman had witnessed the slaughter of animals and fish, he always had the same thought: in their behavior towards creatures, all men were Nazis. The smugness with which man could do with other species as he pleased exemplified the most extreme racist theories, the principle that might is right."

Isaac Singer also famously said that animals, especially animals raised for food, endure "an eternal Treblinka" at the hands of human beings, a reference to the Jewish extermination camp that the Nazis operated in Poland during the Second World War. The language of oppression is deployed to support institutionalized speciesism, where non-human animals become objectified as "it"s rather than "he"s or "she"s with distinct identities. If we were publicly exposed, maiming, raping, enslaving or murdering another human being, we would go to prison in almost every country on Earth. But if we maim, rape, enslave or kill an animal, we could still be regarded as fine, upstanding members of society in almost every country on Earth. In the United States, taxpayers subsidize practitioners of such behavior[36]. To add insult to the injury of animals, in the US, it is those who actively oppose the maiming, raping, enslaving and killing, who would be subjected to prosecution under the USA Patriot Act and the Animal Enterprise Terrorism Act for impeding the profits of the Animal Agriculture industry[37]. In many US states that have "Ag Gag" laws on the books, it is illegal for citizens to exercise their First Amendment rights to free speech and document the maiming, raping, enslaving and killing of non-human beings[38]! The current socioeconomic system vitally depends upon the perpetuation of violence towards animals to promote consumption as an organizing value.

Can we see perfection in all this violence and oppression? Speciesism is inextricably tied to the concept of property ownership and property rights that allow human beings to fence in land and "tame Nature." How can such sheer arrogance be part of perfection? It is only the rarest of souls, such as Pam and Anil Malhotra of SAI Sanctuary, who acquire property rights to tear down the fences and return land back to Nature.

Further, when we assert that, "Everything is Perfect" now, then everything must have been perfect in the past as well. Therefore, can we see perfection in the Nazi holocaust itself? Or more generally, can we see perfection in the documented genocides of the past, the genocide of Central Asian peoples in Stalin's Russia, the genocide of American Indians in the US, the genocide of Armenians in Turkey, the genocide of Cambodians during Pol Pot's regime, the genocide of Indians and Pakistanis during the Great Migration of 1948? Or, how about the ongoing genocide in the Middle East in the endless Global War on Terrorism? Or the genocide in the Congo, where 6 million people have been killed since the mid 90s, mainly to support the industrial procurement of cheap minerals from that resource rich country, so that we can all have disposable electronics and ever-fancier cell phones[39]?

Can we see perfection in the carnage of the First and Second World Wars, in the instant nuclear holocausts of Hiroshima and Nagasaki? Can we see perfection in nuclear accidents such as Chernobyl and in the ongoing, slow motion catastrophe of Fukushima where nuclear cores are still in active meltdown[40]? With respect to humanity's continuing flirtation with nuclear catastrophes, General Lee Butler of the US Strategic Air Command said[41],

> "Humanity has so far survived the nuclear age by some combination of skill, luck and divine intervention, and I suspect the latter in greatest proportion."

The harrowing tales of our lucky nuclear near misses are enough to turn any open-minded atheist into a devout believer in divine Providence.

Can we see perfection in the thousand-fold increase in species extinction rates on Earth? According to the American Museum of

Natural History, we are in the midst of the Sixth Great Mass Extinction event in the Earth's history and this time, it is not due to asteroids, comets or volcanic eruptions, but due to human activities[42]. Species are dying out due to habitat destruction, the introduction of invasive species, chemical pollution, overexploitation, anthropogenic climate change and just plain human overconsumption. While extinction means that the last of the species has died out, almost every surviving species on Earth, except for our domesticated species, has had to endure a catastrophic reduction in numbers. Populations of tigers, lions, giraffes, elephants, dolphins, whales, tuna and salmon, to name a few, are all down 90% or more, though they are not yet extinct. How can we see perfection in all this violence and killing?

Can we see perfection in the relentless progression of climate change? There is no question that the Earth's climate is undergoing rapid changes, with the imminent melting of the sea ice in the Arctic ocean, the irreversible melting of the West Antarctic ice sheet, the potentially irreversible melting of most of the Greenland ice sheet and the continued melting of other land-locked glaciers as lead indicators of such climate change[43]. This melting of the great ice masses on Earth is a nonlinear phenomenon that exhibits hysteresis and can be reversed only if we manage to cool the Earth to pre-industrial levels[44]. As we prepare for coastal cities to be inundated by the rising ocean, we know that some cities such as Miami in Florida, USA, cannot even be saved through the engineering of sea walls etc., as they are situated on a bedrock of porous limestone and the ocean would inundate the cities from below. Further, with increasing temperatures, we will experience unprecedented extremes in wildfires, droughts and deluges throughout the Earth. How can we see perfection in all this carnage and destruction?

Finally, can we see perfection in the burgeoning human population, which is still increasing towards an estimated peak of 9-10 billion people by 2050? In his book, *Harvesting the Biosphere*, Prof. Vaclav Smil estimates that the dry biomass of all wild land mammals on the planet is 5M tons, while the dry biomass of human beings is 55M tons and the dry biomass of all human domesticated mammals is 120M tons[45]. That is, wild mammals comprise around 3% of the biomass of all land mammals on Earth with humans and their domesticated mammals comprising around 97% of the biomass! According to the UN IPCC AR5, Chapter 11, native ecosystems have now been reduced to a mere 8% of the Earth's land surface area, while the remaining 92% show extensive transmutations from human terra-forming activities. Yet, human encroachment into the habitats of the wild animals continues as human population and consumer appetites continue to grow. Can we see perfection in this truly gargantuan scale of our current human enterprise?

## 3.4 Connecting the Dots

Our story rationalizes most of these seeming imperfections as the necessary suffering that had to be endured in our Caterpillar phase as we built out the technological foundations of our Butterfly phase. It does so by reasoning backwards. During his celebrated commencement speech at Stanford University, the late Steve Jobs, the co-founder of Apple Computer, said[46],

"You can't connect the dots looking forward. You can only connect them looking backwards. So you have to trust that the dots will somehow connect in your future.... Believing that the dots will somehow connect down the road will give you the confidence to follow your heart even when it leads you off the well-worn path."

He was channeling the German philosopher, Johann Goethe, who said[47],

"Life can only be understood backwards; but must be lived forwards."

But this is the basis of faith as well. A sacred Hadith in Islam says[48],

"There are so many merciful signs of Allah behind every misfortune that they surpass the pains and agonies caused by that misfortune."

Our story identifies the patterns that we can discern looking backwards into the past, in order to rationalize most of these seeming imperfections. For any past or present imperfections that still remain to be explained, our story rests on the realistic belief that everything will indeed make sense when we consider them backwards at some future point in time, even if those events might seem inexplicable at the moment. At that future time, we can tell a more complete story of humans, consistent with a more complete story of the Earth, weaving them both together into a cohesive whole, all based on the facts and science, as our understanding evolves.

As Anthony DeMello wrote in his book, *Awareness*[49],

"The trouble with people is that they are busy fixing things that they don't even understand... It never strikes us that things don't need to be fixed... They need to be understood. If we understood them, they'd change. Do you want to change the world? How about beginning with yourself?... Through observation, through understanding, with no interference or judgement on your part.

What you judge, you cannot understand... Observe without a desire to change what is. Because, if you desire to change what is into what you think should be, you no longer understand... The day you attain a posture like that, you will experience a miracle. You will change - effortlessly, correctly. Change will happen, you will not have to bring it about. As the light of awareness settles upon the darkness, whatever is evil will disappear. Whatever is good will be fostered."

Clearly, in DeMello's view, to awaken is to become truly aware of the perfection of the present. For it is only when we see everything as perfect that we could "observe without a desire to change what is." With that awareness comes understanding and with that understanding comes right decisions about present and future actions. To attain such awareness, in the Upanishads, it is said that we need to correctly answer the three fundamental questions of the Universe:

1. Who are you?
2. What is your relationship with the world?
3. Why are you here?

Every thought, word or deed of every human being contains implicit or explicit answers to these three fundamental questions. Our story considers these three questions within a scientific framework at the level of our species. That is, as a species, we need to answer:

1. Who are we?
2. What is our relationship with the world?
3. Why are we here?

While answering these questions, our story places humanity as a species that is neither superior, nor inferior, but on par with other species such as the elephant herds of the Western Ghats of India, while playing an equally vital role in the tapestry of all Life on Earth. It is a story that closely aligns with all the faith and wisdom traditions of the world, including atheism in the form of secular humanism.

# 4. Who Are We?

*"The historical mission of our times is to reinvent the human —at the species level, with critical reflection, within the community of life-systems"* - Father Thomas Berry.

In general, we orient ourselves based on a creation story, a story that places us within the Universe and imbues our life with meaning and value. At moments of crises, we tend to question our creation story, if it ceases to explain the changed circumstances. In his recently published monograph, *A New Story for a New Economy*, David Korten outlined four contrasting creation stories conveying very different understandings of relationships, agency and meaning for human beings and our place in the Universe[1]:

"**1. Distant Patriarch:** My most important relationship is to a distant God who is Creation's sole source of agency and meaning.

**2. Grand Machine:** I exist in a mechanistically interconnected cosmos devoid of agency and possessing no purpose or meaning.

**3. Mystical Unity:** Relationships, agency, and meaning are all artifacts of the illusion of separation; I am one with the timeless eternal One.

**4. Living Universe:** I am an intelligent, self-directing participant in a conscious, interconnected self-organizing cosmos on a journey of self-discovery toward ever-greater complexity, beauty, awareness, and possibility."

David Korten proposed that the Living Universe story become the common story that we must all embrace for our future evolution as a species.

However, in the Hindu perspective, all four of these creation stories are valid depending upon the spiritual orientation of the individual human being. In a dialogue related in the Hindu epic, Ramayana, God in the incarnation as Lord Rama, asks his devotee, Hanuman[2],

"What do you think of me?"

And Hanuman replies,

"As this body, I am your servant (Distant Patriarch Story or the **Dwaita** philosophy of Madvacharya[3]).

As this mind, I am a part of you (Living Universe Story or the **Vishishtadvaita** philosophy of Ramanuja[4]).

As this spirit, you and I are the same (Mystical Unity Story or the **Advaita** philosophy of Adi Shankara[5])."

This is how the ancient Hindus reconciled different creation stories and incorporated them within their own lives. Indeed, since the Hindu Brahman is defined to animate each and every particle in the Universe, the Grand Machine creation story has validity as well. Now, even neuroscientists have confirmed that our thoughts and actions are not born of free will, just as predicted in the Grand Machine story[6].

## 4.1 Our Disconnected Lives

Therefore, the difficulty is not with the creation stories per se, but with the cultural stories that rule our daily lives. When our cultural stories don't match reality, we tend to do things that are not true to ourselves, that are against our core "Dharma", Dharma is an ancient Sanskrit word with no precise English equivalent, but which can be loosely translated as "right action" or "action in concert with Nature." It is only through science that we have begun to discover how our cultural stories are fundamentally contradicting reality and therefore, contrary to whom we truly are.

At the moment, we appear confused about who we are as a species, especially within our modern technological civilization, despite the fact that we have been going through a self-absorbed phase in which we seem truly narcissistic. In any university, apart from a couple of underfunded departments of zoology and the environment, we spend all our time and effort studying ourselves and our systems, though in Nature, there are 10-100 million non-human species as opposed to our one species. Despite this self-absorption, we still haven't satisfactorily identified our ecological niche, a unique place within ecosystems where we, as a species, belong exactly as we are. Because we are so unsure of this identity as a species, we don't generally have a sense of belonging in Nature.

If you are reading this in our modern technological world, please look around you. The odds are that every object in your vicinity was made by humans somewhere. It is very likely that you don't understand how these objects were made either, as we have become so disconnected from the industrial processes of production. Even the experts among us are narrowly oriented. There are very few of us who are trained to take a broad systems perspective. As the author and historian, Prof. James McWilliams, wrote[7],

> "Imagine living in the 18th century. Almost everything about your physical existence would make immediate and intuitive sense. Your food, your shoes, your clothes, your transportation, your garden, the mill that churned your flour, your house---these would hold few mysteries in terms of how they came to be and how they operated. Spiritual conundrums might haunt you. But not the logistics of the physical world. It was all levers and pulleys and other manifestations of forces visible.
>
> Now imagine the physicality of your existence today. Can you really explain how your iPhone works? Email? Do I have any idea how this post will appear in hundreds of inboxes of people I don't know? How does an elevator operate? A car engine? The cloud? The bomb? My toilet? The gun that killed Mike Brown?

It's safe to say that at some point in the twentieth century modern humans went from engaging with the physical world from a position of understanding to a position of trust. Blind trust. The first books I ever read were called Tell Me Why, but I remain essentially clueless about the inner mechanisms of the objects that surround me. Every day I ask "why," shrug my shoulders, send my emails, grab the wheel, and view the details of my physical life as comprehensible as Chinese algebra. I just stand back and marvel at it. Or I just hit send."

This disconnection has served to separate the consumer from the violence that was embedded in the production process. It is only the global elites, the billionaires and the business CEOs who have to deal with the slave labor, the suicide nets in factories, the dismembering and maiming of human and animal bodies in slaughterhouses, the CIA-sponsored coups and the brutal dictators in distant countries who help keep our product prices low. They are the hardened few whose empathic cores have been squeezed out through repeated exposure to the violence. The rest of us would be truly horrified if we became aware of the suffering embedded in the products that we consume.

The complexity of our human enterprise has clearly contributed to this disconnection as well, since we have each become more and more silo'ed in our expertise. But even among experts on the topic, opinions differ as to who we are as a species. Some call us a virus, a plague, an evolutionary cul-de-sac, or a failed mutation that Nature will eventually need to eradicate and start over. Others call us the crown of all Creation, a deified species that will eventually replace the native ecosystems of the planet with genetically engineered versions and remake the Earth to our own liking. The truth undoubtedly lies somewhere between these two extremes. The truth is we are just like any other species!

## 4.2 Our Ecological Niche

It is necessary for us to correctly answer the *"Who Are We?"* question, at least from a purely biological sense so that we can put the puzzle pieces together on the much more important *"Why Are*

*We Here?"* question later on. Fortunately, this task isn't so difficult if we focus on what makes our species unique biologically. That is, what are the unique skills that we possess which can be deployed to help the planet's ecosystems thrive? Biologically, we know that we are close cousins to primates and that we share a substantial portion of our genetic code with almost every complex life form on the planet. We know that we are an agglomeration of human cells but our bodies also house ten times as many foreign microbial cells as human cells. Since the mitochondria in human cells are bacterial in origin, each of us is literally a cooperative microbial colony that experiences an integral consciousness and agency.

We also know who we are in terms of sense perceptions and physical abilities. We are really quite an ordinary mammal species; we don't see too well, we don't hear too well, we don't smell too well, we don't climb too well and we don't run too well. Other non-human beings have us handily beat in each of these sensory and motor skills.

Eagles, hawks and buzzards have a sense of sight that is 3-4 times sharper than ours. A buzzard can track a small rodent from a height of 15000 feet and dive at 100mph while keeping the rodent in constant focus[8].

The Greater Wax Moth has a sense of hearing that ranges 15 times higher than ours. While humans can hear sounds up to a frequency of 20KHz, the moth can hear sounds up to a frequency of 300KHz! This enables the moth to hear a bat's sonar signals and thus, escape from becoming the bat's breakfast[9].

Bloodhounds have a sense of smell that is ten million to one hundred million times more sensitive than our sense of smell. Bears have a sense of smell that is seven times more acute than that of bloodhounds[10]!

The Alpine Ibex can easily climb up near vertical surfaces for grazing and for evading predators, while we have a hard time emulating even our close cousins, the bonobos, in climbing up trees[11].

The peregrine falcon can travel at speeds of 240mph, almost nine times faster than Usain Bolt's top recorded speed of 28mph. Among land animals, the cheetah clocks in at over 60mph, the lion at over 50mph, the tiger and the hyena at around 40mph, and all of them are routinely much faster than the fastest human ever and therefore, perfectly capable of capturing Mr. Bolt and turning him into a meal if he was not protected by the technological defenses of our human civilization[12].

The early days of our human ancestors must have been spent in constant fear of predators. We were strong only when we were together, but when alone, we were easy prey. For we, humans, were easy to catch, didn't have claws and were quite defenseless until we learnt to fashion spears and other weapons to help us fight predators from a distance. Indeed, it is now known that the matrilineal most recent common ancestor of all currently living human beings, the Mitochondrial Eve, is only 100,000-200,000 years old, which means that the descendants of all other female contemporaries of our Mitochondrial Eve did not survive into the present[13]. Therefore, it must have been a really tough, terror-stricken life for our early human ancestors.

Thus our human species was born into fear within an environment that was literally red in tooth and claw. It was fear and the quest for survival that shaped early human behavior and which has been coded into our cultural memory through our ancestors. From the very beginning, fear, specifically fear of death, was our default emotional state with love and compassion as an exception, instead of the other way around. But all that we now consider to be evil is rooted in this same fear for we have been undergoing a steady transition away from this primordial fear towards love and compassion.

The principles of Yoga assert that fear is the root emotion for our disconnection from our true selves. It traps our life energies in our base "chakra". That in order for us to become truly successful in our pursuit of happiness, we must let go of fear, guilt, shame, grief, lies, illusions and earthly attachments, in that order, as we ascend in our spiritual development towards enlightenment, our personal moral singularity. To accomplish that, it helps not to be ensconced in a socioeconomic system that is fundamentally rooted in fear!

But our birth in such a fearful environment was necessary to help hone the intelligence of our species. The survivors of the pre-civilizational carnage must have become self-selected for intelligence, for it is only the most intelligent of our species that could have escaped the fearsome predators of that early era. We see the same phenomenon occurring in other non-human beings today. From the hunted, we humans have become the hunters, killing coyotes, wolves, bears, mountain lions and other predators en masse in order to protect our expansive livestock herds. In the process, the wild animals and especially, coyotes, have become self-selected to be incredibly intelligent to the point where they have been consistently outwitting their human hunters over the past few years. Coyote hunting has become a specialist profession and many coyote hunters will attest that if coyotes had opposable thumbs, they would be ruling the world today[14]!

Not only did we become self-selected for intelligence and learned to harness fire, we also formed partnerships with dogs about 50,000 years ago, who lent us their superior sense of smell and hearing in exchange for food[15]. This partnership enhanced the probability of survival of the human species, for a human with a canine companion is far harder to prey upon than a human alone. We also developed an expanded vocabulary of language to communicate with each other, mainly to describe the wider range of foods that were available to us through cooking with fire. But language is a skill that is common to all social species, not just humans. Dolphins have sophisticated languages and so do prairie dogs, lizards, birds and most other non-human beings[16].

Prof. Con Slobodchikoff, Professor Emeritus of Biology at Northern Arizona University in Flagstaff, Arizona, has decoded the language of the prairie dog systematically by recording the sounds emitted by the prairie dog sentinel of a prairie dog colony in response to various stimuli. At this point, it is the most sophisticated non-human language ever decoded so far, and it is breathtakingly complex and amazingly efficient[17]. Prof.

Slobodchikoff believes that sometime in the future, humans will appreciate the language of other social species and totally understand what they are telling each other and telling us and respond back to them in their own language. Through his careful scientific experiments, he has shown that many non-human beings describe the world around them with words and gestures making language skills a continuum among all species and not just an exception, unique to humans.

If language skills are a continuum among all species, then so is the accumulation of knowledge over generations. This is aptly illustrated by the behavior of elephant herds, flocks of birds and even butterflies and other insects that clearly pass on knowledge of water holes and migration paths from generation to generation.

Therefore, what is unique about humans? What defines us? David Korten said[18],

> "We humans are born with a capacity distinctive among Earth's species to reflect on our own mortality, ponder the meaning of Creation, and ask "Why?" By our answers, we define ourselves, our possibilities, and our place in the cosmos."

But the trouble is that we haven't yet satisfactorily answered the "Why?" question. Indeed, if anything, we humans are distinctive among Earth's species, not for knowing our place in the cosmos, but for not knowing it. The elephant herds of the Western Ghats have us handily beat in that regard. Besides, we can't really tell what the elephant herds in the Western Ghats are truly thinking, either. What if they have already answered David Korten's "Why?" question since they clearly know their place in forest ecosystems and are now asking a different "Why?" question, as in,

> "Why can't humans stop using our tusks to make trinkets and leave us in peace so that we fulfill our possibilities, and our place in the cosmos?"

Indeed, Charles Darwin wrote about 150 years ago that[19],

"The difference in mind between man and the higher animals, great as it is, certainly is one of degree and not of kind."

But despite Darwin's observation, in the Western perspective, it had remained axiomatic that only humans were conscious beings, while all other non-human beings were assumed to be mechanical, emotionless, unconscious beings, incapable of feeling pain, pleasure or anxieties about the future. But animal behavioral scientists have been knocking down each of these assumptions one by one. Anyone who has ever lived with a dog or a cat would know that they are perfectly capable of emotions, feel pleasure and pain, anxiety and joy. But science depended upon such a mechanistic model of non-human beings in order to justify the live-animal experiments that were necessary for early scientific progress. To this day, many conscientious scientists chafe under the requirement for mandatory live-animal experiments needed for the approval of medical treatments and procedures by government regulatory authorities, for they suspect that those animals being experimented upon have feelings as well. Finally, after centuries of mostly willful ignorance on this issue, a prominent group of cognitive neuroscientists, neuropharmacologists, neurophysiologists, neuroanatomists and computational neuroscientists signed the following Cambridge Declaration on Consciousness on July 7, 2012[20]:

"The absence of a neocortex does not appear to preclude an organism from experiencing affective states. Convergent evidence indicates that non-human animals have the neuroanatomical, neurochemical, and neurophysiological substrates of conscious states along with the capacity to exhibit intentional behaviors. Consequently, the weight of evidence indicates that humans are not unique in possessing the neurological substrates that generate consciousness. Nonhuman animals, including all mammals and birds, and many other creatures, including octopuses, also possess these neurological substrates."

While the Cambridge Declaration on Consciousness was sweeping, it didn't specifically ascribe consciousness to fishes, crustaceans and insects. But that is likely to change soon. In a recent paper authored by French scientists, it was demonstrated that crayfish exhibit a form of anxiety similar to that described in vertebrates, suggesting the conservation of several underlying mechanisms during evolution[21]. Another recent paper published in the journal, Animal Cognition, showed that[22],

> "Fish perception and cognition match or exceed that of other vertebrates... fish compare favorably to humans and other terrestrial vertebrates across a range of intelligence tests."

Western and scientific perspectives are thus becoming more in alignment with dominant Eastern and indigenous perspectives, where it had been accepted for millennia that all Life is conscious from the smallest microbe to the largest of animals. All life forms have evolved to their present state over billions of years and as such, they are all contemporaries of our species on an equal footing. In the Upanishads, it is said that consciousness is in everything and everything is in our consciousness. The Buddha stated 2500 years ago[23],

> "All beings tremble before violence. All fear death. All love life. See yourself in others. Then whom can you hurt? What harm can you do?"

This acceptance of the consciousness of all Life birthed the concept of Ahimsa, compassion for all Creation or kindness to all Life, in India, a few millennia ago[24]. Therefore, consciousness, far from being unique to humans, appears to be universal and science is beginning to accept that, though grudgingly.

From a purely mechanical perspective, we do know that humans are unique tool builders. While other non-human beings such as crows and higher primates build tools and use them, our controlled use of fire and our opposable thumbs help us to be superlative tool builders. It is the fear of predators that initially drove us to be great tool builders for we needed the spears, the javelins, the bows and arrows in order to fight our predators from a safe distance. Later, it was the fear of other humans that led us to continue developing this tool-making capability for militaristic purposes. As Steve Jobs pointed out[25],

> "I think one of the things that really separates us from the high primates is that we're tool builders. I read a study that measured the efficiency of locomotion for various species on the planet. The condor used the least energy to move a kilometer. And, humans came in with a rather unimpressive showing, about a third of the way down the list. It was not too proud a showing for the crown of creation. So, that didn't look so good. But, then somebody at Scientific American had the insight to test the efficiency of locomotion for a man on a bicycle. And, a man on a bicycle, a human on a bicycle, blew the condor away, completely off the top of the charts."

With our tools and technologies, we are able to enhance all our senses to be superior to those of any other species on Earth and thereby, outwit, outpace and outcompete them all. This is how humans came to dominate the Earth and consider themselves "the crown of creation." The birth of our species in an environment of abject terror helped hone our superlative tool building skills. Conversely, our superlative tool building skills has helped us successfully shed all fear of our natural predators, except predators of our own kind. Nevertheless, we continue to admire the predators of the world, the tigers, the lions and the cheetahs. As any safari owner in Africa would tell you, it is the carnivores that attract the tourists, not the herbivores. We admire these predators and strive to emulate them, especially in our diets. It is to support our diets that we perpetrate the vast majority of the institutionalized, deliberate violence towards other beings in our human societies[26].

## 4.3 Our Vegan Conundrum

Unfortunately, humans are not biologically equipped to be like carnivores. Our biological characteristics such as our colons, intestines, saliva, jaws and teeth, which have been relatively unchanged for the past 200,000 years, are not designed for the consumption of animal flesh. It is only through cooking with fire that we make animal flesh edible and digestible for our bodies, but even this causes harmful changes in our livers and in our digestive systems. Therefore, prior to the discovery of the controlled use of fire, it is safe to say that our ancestors subsisted primarily on foraged plant-foods and perhaps, small animals and insects, just as chimpanzees and bonobos do today. Says David Nibert, a Professor of Sociology at Wittenberg University[27],

> "Few people are aware that, for most of our time on the planet, our species foraged and lived primarily on plant-based diets. Our communities were egalitarian and there was ample time for leisure and social activities. This long period has been referred to by anthropologists as "the original affluent society." However, this era ended when humans began routinely to hunt large animals – primarily a male pursuit. As our species does not have the biological make-up of a predator, this hunting could only be accomplished through the creation of weapons. Those men most successful at such killing exerted growing power. Social hierarchy began to emerge, and the status of women began to decline."

It is the controlled use of fire that allowed humans to expand their range from out of Africa to span the entire globe. Fire extended the foods available for human consumption, especially meat, which made it possible for humans to survive even in cold, harsh climates. During the past 200,000 years, the Earth's climate has undergone several dramatic changes, swinging from ice ages to warm interglacial periods in a few thousand years, a geological blink of an eye. But over the past 10,000 years, the Earth's climate has been relatively stable, spawning the agricultural revolution and the controlled production of grain crops with the help of domesticated animals. As pockets of stable, steady state human civilizations grew in various parts of the globe, it is the Westerners who had migrated to the cold, harsh climates who developed the tools and science-based military technologies necessary to conquer and forcibly reunite the human family globally during the last few centuries.

Since the conquering class consumed plenty of animal foods, such consumption has now become a sign of upward mobility globally. But there is considerable angst evident in our decisions to consume such foods. Fundamentally, though our species was baptized in fear, we possess the capacity for the dual emotions of both fear and love, just like most other complex species. We are born with a capacity for fear in order to escape predators and with a capacity for love in order to nurture our young. Our long 200,000 year journey from the savannas of Africa to our technologically connected global presence today has been a journey from fear as our predominant emotion towards love and compassion as our predominant emotion. This is what Rev. Martin Luther King, Jr., alluded to when he said[28],

> "The arc of the moral universe is long, but it bends towards justice."

But remnants of our past create mass cognitive dissonance within us along our journey. We eat animal foods, but we tell ourselves stories about such choices that no longer match reality. We had formed partnerships with some animal species, such as dogs, but now we treat them as family members in most countries around the world, while we skin them, cook them and eat them in the Far East. Anyone who has lived with dogs as pets would empathize with the beseeching eyes of the dogs in the Far East as they are being carted off for slaughter, packed in the dozens to a single bicycle carrier. We wonder why the consumers in the Far East cannot see these same eyes of the dogs and not be moved to tears. Yet we fail to see the same emotions in the eyes of the cows, pigs and chickens that we exploit and slaughter en masse in other parts of the world.

Compassion is at the very core of our being. Would you ever deliberately hurt an innocent animal unnecessarily? So far, among the thousands of people that I have asked this simple question, not a single person has come forward to say, "Yes." Of course, I have probably never spoken to trophy hunters like Donald Trump, Jr., who famously cuts off an elephant's tail and poses with it after shooting the elephant dead. Or perhaps, the trophy hunters among my responders were ashamed of their hobby! In either case, this goes to show that compassion for all Creation is coded into every fiber of our human being.

That's who we really are! Therefore, is it any wonder that many religious texts of the world proclaim that Man was created in the image of God? After all, God is commonly defined to be all compassionate, in addition to being omnipresent, omniscient and omnipotent. God is Love. And so is Man, except that Man's ego can blind him to that truth. But every human being has the potential to realize that truth as a lived reality. That is our destiny on the road to sustainability.

But among those same responders to my question, very few admit to being vegan, even though by definition,

**Veganism is a way of living where we seek to never deliberately hurt an innocent animal unnecessarily.**

Many people, especially in the older generations, feel repulsed by that word even though its meaning is part of their identity. That is because the word "vegan" has become closely associated with diet and many of those who would "never deliberately hurt an innocent animal unnecessarily," do continue to eat meat, fish, dairy and eggs. They were raised on cultural stories that the consumption of these animal foods is necessary for human well-being, even though the scientific evidence is now overwhelming that this is false.

Members of the American Dietetic Association and the Canadian Dietetic Association wrote a scientific paper recently clearly stating that it is now unnecessary to eat animal foods of any kind at any stage of our human life cycle[29]. Yet, animals that had been domesticated over thousands of years to provide muscle power for our ploughs and manure for our fields, are now being raised strictly to provide meat and dairy for our consumption and raw material for our clothing. These animals are now treated the equivalent of crops, as they are forcibly reproduced, raised and harvested in their youthful prime, in giant industrial operations. We were told stories to justify the institutionalized deliberate violence that is modern animal husbandry. Some of us were told that the animals lived a good life grazing on pastures and then experienced one bad day when they were slaughtered for meat. Some of us were told that the animals themselves offered up their bodies, even going so far as to place their necks on the butchers' knives, without any coercion. Some of us were told that eating animals is normal, for that is what predators do. It would be natural as well, so long as we thank the animals for giving their lives and we consume all parts of the animals without any wastage. We suspect that these stories are false, especially when we become aware of factory farms and feedlots, but we soldier on with our consumption regardless. In my case, I was raised a lactovegetarian and I was told when I was young that we take milk for human consumption from the mother cows only after their calves had finished drinking. That is the "Ahimsa" way, where nobody got hurt. In fact, I was told that we were only consuming the excess milk that we anyway had to drain from the udders, for otherwise the mother cows would suffer from mastitis. Therefore, drinking the cow's milk was really an act of compassion, not exploitation!

I really, really wanted to believe that!

## 4.4 My Vegan Conversion

When I immigrated to the US and came to know of the horrific conditions in which dairy cows are raised in the US, I justified my continued consumption of dairy products telling myself that it is different in India. I spun a cocoon of denial around myself so that I didn't have to face the reality of what I was consuming on a daily basis. But the truth kept intruding, time and again.

The dairy industry in the US is one of the most blatantly violent industries in the world, separating mother cows from their calves within days of the calves' birth, torturing the mothers thrice a day, every day, by sucking milk out of their udders in 1 minute flat using machines, forcibly impregnating the mothers again and again and then slaughtering them and grinding them into hamburgers when they are 5 years old, just a quarter of their normal life span. The machine-based milking meant that a certain number of sick cows would necessarily be milked every day and therefore, the US

Food and Drug Administration began to accept a certain percentage of pus and blood in milk on a routine basis. As Dr. Michael Greger puts it[30],

> "According to the USDA, 1 in 6 dairy cows in the United States suffers from clinical mastitis, which is responsible for 1 in 6 dairy cow deaths on U.S. dairy farms. This level of disease is reflected in the concentration of somatic cells in the American milk supply. Somatic cell counts greater than a million per teaspoon are abnormal and "almost always" caused by mastitis. When a cow is infected, greater than 90% of the somatic cells in her milk are neutrophils, the inflammatory immune cells that form pus. The average somatic cell count in U.S. milk per spoonful is 1,120,000...
>
> A study published in the Journal of Dairy Science found that cheese made from high somatic cell count milk had both texture and flavor defects as well as increased clotting time compared to milk conforming to the much more stringent European standards. The U.S. dairy industry, however, insists that there is no food safety risk. If the udders of our factory-farmed dairy cows are inflamed and infected, industry folks say, it doesn't matter, because we pasteurize—the pus gets cooked. But just as parents may not want to feed their children fecal matter in meat even if it's irradiated fecal matter, they might not want to feed their children pasteurized pus."

Moreover, with all that hormonally induced overproduction of milk, cows can barely stand up after 3-4 lactation periods and become "spent". When they are just 4-5 years old, a mere quarter of their normal lifespans, these "spent" dairy cows are then carted off for slaughter and turned into hamburger meat, while all their male offsprings and a substantial fraction of their female offsprings are suitably starved of iron and other essential nutrients and turned into tender veal after a brief, tortured life.

But within my comfortable cocoon, I tried not to think about the plight of the mother cows that were providing me with milk, cheese, yoghurt, ice cream, butter, ghee and above all, those delicious pedas, gulab jamuns and rasgollas, popular milk sweets in Indian cuisine. I tried not to think about the calves that were separated from their mothers within days of birth, telling myself that people don't separate calves from their mothers in India. I tried not to think about all the calves being tortured for veal. After all, in India, cows are revered as sacred and we would never hurt them, would we? Isn't that why dairy is still consumed in India by us, whose ancestors pioneered the concept of Ahimsa? Aren't dairy products, especially clarified butter (ghee), essential to Ayurveda, the ancient medicinal science of India, and essential to numerous sacred Vedic rituals in Hinduism?

But the stories that we tell ourselves even as we have a sinking suspicion that we're fooling ourselves, take a toll on our mental, physical and spiritual health. We are what we eat is not just a cliché, but also a truism. What we eat is a more accurate reflection of who we are than anything else. Patricia Bragg had said[31],

"We are what we eat, drink, think, say and do."

In that line, what we eat comes first. When what we eat, drink and do are incompatible with what we think and say, we suffer tremendously. But thankfully, during a fateful trip to India in December of 2008, I was forced to excavate every one of these cultural stories from the recesses of my mind and reconcile them with reality and with my core values. And make some life-changing decisions!

This fateful trip actually originated three years before, in December of 2005, when I was inspired by Vice President Al Gore's presentation to work on the environment instead of continuing my engineering career. Then I wrote to Mr. Gore and asked how I could be of help to him and he kindly invited me to Nashville, Tennessee, to get trained to make his presentation during December of 2006. This is how I became a member of his Climate Reality Project[32] and went around making presentations at churches, schools and professional gatherings in the US and India. Becoming an engineer trains you to be a doer and therefore, I had also started a non-profit organization called Climate Healers[33] with a mission to truly HEAL the Earth's climate - as opposed to maintaining it precariously in an advanced state of disrepair - and an objective to work with in-country Non Governmental Organizations (NGOs) in India on reforestation to sequester the excess carbon in the Earth's atmosphere. But I continued making presentations on behalf of the Climate Reality Project even after starting this non-profit. Of course, I tailored Mr. Gore's presentation to make it more personal and as part of this customization, I had been showing before and after photographs of a 250-acre protected forest that the villagers of Karech in Rajasthan, India, had nourished over a 4 year period from 2002 to 2006. These photographs showed barren land with a spindly looking tree in the foreground turning lush green with that same unmistakable tree blossoming in the foreground just four years later. During many of these presentations, some audience member would ask whether I was sure that these photographs were taken four years apart and not during different seasons of the same year. What if the before photograph, supposedly from 2002, had really been taken during the heat of summer while the after photograph from 2006 had been taken during winter? What if the so-called "forest regeneration" was just an elaborate scam that the villagers and the Indian NGO, the venerable Foundation for Ecological Security (FES)[34], who provided these photos, were playing on me? Therefore, when I found myself in the village of Karech during that trip in December of 2008, I asked the villagers if they would take me to their protected forest so that I could see it for myself.

They did. What I saw there shook me to the core of my being. Yes, the protected forest was even more beautiful than the 2006 photo that FES had sent me, but on the other side of the fence enclosing the protected forest, the land looked just as barren as the 2002 photo. The protected forest was fenced to prevent livestock from grazing in there, while livestock herds were freely roaming in the unprotected land. It struck me then that it was my consumption of dairy that was responsible for the devastation in the unprotected land.

India has a substantial population of about 600 million ethical lacto-vegetarians who consider the cow to be sacred, but consume milk and milk products on a daily basis. The result is that the cows are constantly impregnated to promote lactation, but they and their babies are not killed as ruthlessly as in the US, leading to a massive bovine population explosion. While the average dairy cow in the US lives for no more than 5 years, cows in Nature live for an average of 20 plus years. As a result, India has over 320 million heads of cattle, more than thrice as much as the US with 90 million heads of cattle, on about one-third the land area of the US. Consequently, the tiger habitat is now down to less than 2% of India's land area, when it used to be 90% just 100 years ago. The forest is dying, the sambhar deer is dying, the elephant is dying and the tiger is dying just to make room for all those cows and buffaloes. Lately, India has been increasing the slaughter rate for cows and buffaloes to keep the bovine population in some check. As a result, India has now become one of the largest exporters of beef in the world surpassing Brazil, with a 25% market share as of 2015, a dubious distinction for the land of Ahimsa[35].

So much for the sacred cow!

Due to the blatant inequities in our monetary system, the villagers of Karech are poverty-stricken and naturally have to do what it takes to survive. If the demand for dairy and meat products is rising along with the burgeoning middle class in India, then the villagers have to supply that demand in order to augment their meager incomes. Just as the British colonial state built railroads to efficiently extract timber from the forests of India in the 19th century, the post-colonial Indian nation state built roads into remote villages to efficiently extract firewood, meat, dairy and timber from the forests of India in the late 20th and early 21st centuries. Banks give villagers loans to buy cows when they can attest that they reside close to wildlife sanctuaries and therefore have access to good grazing grounds. The banks know that even though livestock grazing is not generally allowed in the protected forests, the tribal villagers have grandfathered rights. Over time, with soaring consumer demand for meat and dairy, the land becomes overgrazed. At which point, banks are still willing to give the villagers loans to raise goats since goats eat almost anything, including the roots under ground. When goats get done with the land, it becomes a desert, eventually forcing villagers to move into slums in cities and eke out a living as manual laborers. Of late, the desertification of forested lands has been growing substantially in the state of Rajasthan in India, while the state has become the largest producer of mutton and beef for export to the Middle East.

As I stood there naked in my hypocrisy, waves of shame washed over me. When I left the village of Karech during that trip, I knew that I could no longer stay comfortable in my cocoon of denial and had to eliminate my consumption of dairy products right away. Then I visited a colleague, Amala Akkineni, in the city of Hyderabad and she told me that she had just given up dairy consumption and had turned vegan. This was so serendipitous and when I enquired why, she explained that as the head of an Animal Rights organization, she had been asked to certify that the local slaughterhouse met regulations. While she was able to certify that the slaughterhouse met regulations, she saw what was being slaughtered and immediately turned vegan.

I asked,

"So, what was being slaughtered?"

And she replied - I am loosely paraphrasing here -,

"They were buffaloes with shiny skins, who were obviously well taken care of. As they came off the lorries (trucks), they had this quizzical look on their faces because they could see the knives coming down at the end of the line. They were all mothers who had just stopped getting pregnant and this one slaughterhouse was killing 500 of them each and every day! From that day onward, I stopped drinking milk because I'm haunted by the looks on the faces of those buffaloes as they came off the lorries. Now I make yoghurt using soy milk and use that for our curd rice."

Curd rice is a staple in South Indian diets as it has live cultures to reinforce stomach bacteria and promote digestion. Almost every major meal in that part of the world is ended with that dish. I was thrilled to discover that even after quitting dairy consumption, I didn't have to give up my curd rice!

A few days later, I was relating my experiences to a friend and he showed me an online video of buffaloes teaming up to rescue a calf from lions in the Kruger National Park in South Africa[36]. When I watched that clip, I felt sure that every buffalo mother would like to do to me what those buffaloes did to the lions, for stealing their baby's milk. That cemented my decision to quit my dairy habit.

At first, when I quit dairy consumption, I thought I would miss those milk sweets. But a week after I quit consuming dairy, I had this huge sense of guilt lift off my shoulders. I did not know why I had been harboring all that guilt, but the relief was so immense and palpable. I knew then that I would never go back to that destructive habit, ever. Within a month after I quit consuming dairy, I no longer felt even the slightest twinge of the arthritic pains that had been the source of my constant suffering for the previous eight years. Instead, I felt so alive and energetic that I thought I could play cricket once again as well as I did when I was a teenager in India! Back in 2000, when I was diagnosed with arthritis at the age of 40, I became resigned to it as my genetic inheritance from my father who had suffered from arthritis for the last 30 years of his life. Now, I really regret not giving up dairy much sooner when I became aware of the inherent cruelty in the industry in America in the early 90s. To think that I could have helped my father avoid the constant joint pains that wracked him all those years by giving up dairy!

Three years later, during another visit to the same village of Karech, I happened to be watching a woman milking her cow and then the light bulb turned on. I understood why I had been carrying that enormous feeling of guilt over my dairy consumption. I recalled a conversation between my grandmother and my grandfather from when I was about 6 years old. At that time, we were living in the metropolis of Chennai on the East coast of India while my grandparents were living in the village of Nethrakere near Mangalore on the West coast of India. During our summer vacations from school, my parents used to send all the children over to the grandparents' homes to spend some time with them. As children, we really looked forward to these trips since we got to explore in the woods and spend time with our cousins.

One evening, I overheard my grandmother discuss the milk situation with my grandfather. Clearly, my grandparents were still practicing the ancient tradition of milking their cows after the calves had fed their fill. But my grandmother was complaining to my grandfather that this particular male calf was drinking too much and was not leaving enough for the children.

My grandfather told my grandmother to pull the calf away after ten minutes!

I knew then that something wrong was going on, but in the turbulent excitement of childhood, I filed it away in the back of my mind and forgot about it.

Or did I?

I'm now certain that was the primary reason for the guilt that I had been harboring all along, which finally surfaced as relief when I became committed to Veganism. Guilt occurs when we act against our core values, our Dharma. I felt guilty about causing suffering to that poor calf, who was being deliberately deprived of his mother's milk, because compassion for all Creation or kindness to all Life is the core Dharma of our species. Just as the elephants' Dharma is to conduct themselves in ways that make the forest thrive, our Dharma is to nurture Life with our compassion and help all Life thrive. It is no coincidence that virtually every one of the world's wisdom traditions, including every major religion, preaches compassion for all Creation. The underlying truth that they are all trying to convey is actually the core Dharma of our species. We have a strong need to be compassionate, perhaps even more than our need for compassion.

But when we deliberately act contrary to our core Dharma, our minds store the resulting guilt as an imprint in our subconscious which then gnaws away at our spiritual, mental and physical well being. These subconscious imprints are known as "Samskaras" in Sanskrit. How often have we heard people say that the first time they witnessed an atrocity, say the slaughter of an animal, it really affected them badly but then they got used to it? Clearly, it affected them the first time because their inner being was screaming at them to intervene and stop the atrocity, the oppression of an innocent other. Then, assuming that they are powerless to stop the atrocity, their minds use this subconscious imprint, the "Samskara," to dull their senses to the event, if it ever occurs again. This is Nature's mechanism to reduce the instantaneous impact of any continuing atrocities and thereby minimize our suffering, but it doesn't mean that our core Dharma has somehow changed. It certainly doesn't mean that our silent witnessing of the continuing atrocities, or worse yet, participating in them directly or indirectly, is inconsequential to our well-being. The best that the Samskara can do is to dull the instantaneous guilt, not eliminate the accumulated suffering. When our actions don't match our words, we wound our souls.

The ancient sages of India, especially Buddha, who is undoubtedly the greatest scientist of the human mind of all time, understood all this 2500 years ago and devised simple meditation techniques to help us reconnect with our core Dharma. The Buddha was a true scientist in that he conducted observational experiments on himself and then verified that what worked for him also worked for others around him. The Vipassana "insight" meditation technique that the Buddha taught helps us uncover each of these defining, subconsciously stored, personal Samskaras that are clouding our perception of our core Dharma[37]. Once these stored Samskaras are uncovered, they can be acknowledged and rendered harmless through equanimity, provided that we are not continuing to be complicit in the underlying atrocity. Therefore, the conscious adoption of an attitude of compassion towards all Creation, Ahimsa, is an essential first step in our path towards liberation.

Even in the land that birthed the concept of Ahimsa, we have
strayed so far from it in practice! Perhaps, it all begins with
conversations similar to the one between my grandparents, but
there is no doubt that animals including the sacred cow are
exploited cruelly in India as well. The average cow in the cities of
India has been found to have 70 pounds of plastic lodged in her
stomachs as she tries to nurture herself on household waste in the
city streets. The cow leads a miserable life even in the villages of
India, as I witnessed. But the cow is also proliferating needlessly in
India, for dairy products are not necessary for human well-being.
Even the use of milk products in Vedic rituals can be easily
substituted with coconut products for the symbolism carries over
almost one to one.

### 4.5 The Symbolism of Rituals

In the Hindu view, all that is good is God or Brahman. Truth is
God, happiness is God, peace is God, and so on. All evil occurs due
to the human ego, which God overcomes in order to maintain
sanity in the universe. Hindus use idols and rituals to symbolize
various aspects of God and our relationship with God. Milk is used
to symbolize the fluid mind in some Vedic rituals, but so is coconut
water in other rituals. Butter is used to symbolize the solid, steady
mind that is fixed on God, but so is coconut meat. Ghee, which is
clarified butter, is used to symbolize the clarified, steady mind that
is ready to be enlightened (literally lit with a flame), and this can
clearly be replaced with coconut oil. After all, in every temple puja
ceremony, Hindu devotees shatter a coconut to symbolize the
shattering of their ego (the coconut) to release the fluid mind
(coconut water) so that the solid mind fixed on God remains (the
coconut meat), leading to enlightenment (the lit lamp). When we
truly understand the extreme abuse that modern cows routinely go
through, we would refrain from using the products of all that
violence in our sacred rituals to invoke the divine.

The Hindu tradition is full of stories and symbols that helped the
cultural transmission of Dharma, or right conduct, in every
situation. The symbolism of milk and butter comes from the epic
Mahabharata, specifically from the stories of Lord Krishna, who

was raised a cowherd. As a child, Lord Krishna loved butter and even stole it from the households in his neighborhood. Clearly, these stories originated at a time when Indian society was mostly agrarian and depended entirely on animal labor for its well-being, but that doesn't mean that we should consume butter as if we were reenacting Lord Krishna's mythical exploits in modern times.

There is tremendous wisdom encoded in Hindu symbols, but they don't all make sense when taken literally. For instance, before starting any major enterprise, a practicing Hindu will perform a ceremony dedicated to Lord Ganesha, the Elephant God, the remover of obstacles and the symbol of wisdom. Lord Ganesha is depicted as a potbellied man with an elephant head with one broken tusk, four hands holding various objects, typically, an axe, a rope, a conch shell and a sweetmeat. Lord Ganesha also uses a mouse as his vehicle. Clearly, the intent of the ceremony is to imbue the devotee with this image of Ganesha so that she/he attains the right attitude for conducting that enterprise successfully. There are ten aspects of Ganesha's image that pertain to this right attitude:

1. The rope in his hand is to corral the devotee to follow the righteous path of Dharma.

2. The axe is to remind the devotee to cut off all attachments to the fruits of the undertaking.

3. The conch shell is to remind the devotee to take into account the needs of the entire universe, when performing the undertaking. The conch is the symbol of the Universe as it makes the primordial "Om" sound when we blow through it.

4. The sweetmeat held out in his outstretched palm is to remind the devotee to perform the undertaking as a gift to the community.

5. The narrow eyes of the elephant head indicates that the devotee must be far sighted and consider the long term implications of the undertaking.

6. The trunk of the elephant head, which is capable of huge tasks such as uprooting a tree and nimble tasks such as picking up a peanut, indicates that the devotee must consider both the broad view as well as the minute details of the undertaking.

7. The broken tusk of the elephant head is to remind the devotee of her/his own imperfections.

8. The large ears of the elephant indicate that the devotee must listen to the wisdom of others while making decisions during the course of the undertaking.

9. The potbelly is to remind the devotee that she/he also possesses knowledge for the undertaking and need not entirely be reliant on others.

10. Finally, the mouse as a vehicle for Lord Ganesha indicates to the devotee to think outside the box of conventional wisdom, for sometimes the seemingly impossible can be achievable.

Such is the symbolism of Lord Ganesha. However, as the story goes, Lord Ganesha got his elephant head because his father, Lord Shiva, beheaded him in a fit of rage born of mistaken identity and when later struck with remorse, grafted a passing elephant's head on his body instead. But that doesn't mean we should be looking seriously into perfecting the technology of elephant head transplants for all our first-borns. Indeed, the Prime Minister of India, Narendra Modi, was mercilessly mocked in the Indian media for his literal interpretation of the story of Lord Ganesha when he claimed that the story implies ancient Indians must have developed advanced plastic surgery techniques, thousands of years ago[38]. Yet, I'm sure that many of those who were mocking the unfortunate Prime Minister are still consuming the maternal secretions of an entirely different species, the cow, simply because Lord Krishna did so in the Mahabharata!

Truly, every cultural story that we have used to justify our continued violence towards animals has now become fully hollowed out in our modern industrial reality.

Firstly, it is impractical to raise animals in traditional pastures to satisfy the current demand for animal products, as there is not enough land on Earth to meet it.

Secondly, even if we can persuade animals to offer up their bodies willingly for slaughter, that doesn't negate the fact that all the animals we slaughter these days are really babies who have experienced less than 10% of their natural life spans, on average. That act of "consensual slaughter" is therefore just as inappropriate as the act of a pedophile who has consensual sexual relations with a 7-year-old child!

Finally, we can no longer hunt free animals to sustain our current food habits; even if we use every part of the animal we kill, because there are not enough free animals in the wild for us to hunt to sustain our current demands for animal foods. Most certainly, the last remaining forests in the world cannot survive 7.4 billion predatory humans shooting game for their daily meals!

## 4.6 Our Rapid Awakening

When the cultural stories that rule our daily lives don't match reality, that's a sure sign that we have become truly disconnected from our core Dharma. In their book, *The Pleasure Trap: Mastering the Hidden Force that Undermines Health and Happiness*[39]," the psychologists, Doug Lisle and Alan Goldhamer, explore the mental processes that lead to this disconnection and show that other animals can become equally disconnected from their core Dharma when subjected to the same types of stimuli that modern humans experience. The authors begin with the age-old observation that we're born with the sensations of pleasure and displeasure to indicate which of our acts are in harmony with Nature (Dharma) and which acts aren't (Adharma). All beings are infused with the drive to conserve energy, which is to minimize the energy that we expend for each of our actions. Thus an elephant would experience pleasure while eating jackfruits, would experience displeasure if she ever ventured to eat deer flesh and would invariably reach for the nearest available jackfruit in order to conserve her energies instead of seeking one from far away. Likewise, we humans experience pleasure while petting an animal, displeasure while watching an animal being tortured and will reach for the nearest animal to pet, to conserve our energies.

The "Pleasure Trap" arises when we stop using the sensations of pleasure and displeasure as indicator signals from Nature regarding our Dharma and instead use our tremendous tool-building skills to create artificial scenarios, seek out manufactured pleasures and get addicted to them. For instance, processed foods are specifically formulated to maximize our pleasure sensations far beyond that which can be obtained from foods in Nature. As a result, we begin to find Nature's foods, such as fruits and vegetables, to be less pleasurable and consume the processed foods instead. The distortion of our core Dharma and consequent disconnection from Nature intensifies. Over time, this disconnection has become more and more pronounced until many modern humans suffer from isolation, depression and so-called Nature Deficit Disorder[40].

Our story is that this disconnection from Creation occurred deliberately in order to help us develop our tools and technologies and build out a globally connected presence for a global purpose, but as part of an inexorable process towards transformation and reconnection with Creation. Humanity has experienced the Caterpillar phase over the first 200,000 years of its existence and it is currently in the cocoon of disconnection, pondering and awaiting the emergence of the Butterfly phase. Consider any projections of our current course and we will find unbelievable predictions about our near future, meaning that we will inevitably break out of the exponential growth, Caterpillar phase that we're on. For example, if current trends continue unabated, the National Institutes of Health (NIH) is predicting that 3.3 billion people around the world would be overweight or obese in 2030 compared to the 1.3 billion people in 2005, with Americans leading the pack with close to 100% of adults being obese or overweight[41]! If current trends continue unabated, half the children in America are going to be autistic by 2025[42]! If current trends continue unabated, the world economy would be quadruple its current size by 2050, even though it is already 60% larger than the Earth's biological capacity today[43]!

As Paul Gilding points out in his book, *The Great Disruption*[44], all such predictions are truly worthless, since planetary feedback alarm signals such as the climate, extreme weather events, wildfires, famines, floods and droughts will be getting stronger and stronger to force a structural sea-change in the way humans interact with each other and with the Earth. Even if many people willfully ignore these signals today, as time progresses, it will become virtually impossible to ignore them and continue business as usual in the next couple of decades.

Therefore, all the feedback signals from Nature are indicating that the Caterpillar phase of humanity will end within the next decade as we transition from fear of death as the driving force of our civilization to love of life as the driving force. I define the Butterfly era to have begun when half of all Americans self-report to be vegan. By that definition, we can calculate that the Butterfly era will begin within the next decade. A Hartman Group Research report revealed that as of 2010, in the United States, 12% of millennials, born between 1980 and 2000, 4% of Gen-X'ers and 1% of Baby boomers self-reported to be vegan[45]. As to be expected, the younger generation was leading the transformation! Furthermore, since 2010, the Google trend for the search term, "Vegan," has been increasing exponentially and it has tripled to overtake that for a common item like "Coca Cola[46]". The trend is highest in affluent nations such as the United States, Canada, Australia, New Zealand, Germany, Austria and the United Kingdom, implying that this is a lasting trend that will inexorably sweep the world. If that exponential growth trend continues, then within the next 6 years, all Millennials will self-report to be vegan and within the next 12 years, all Gen-X'ers and even 27% of Baby boomers will self-report to be vegan!

In *Carbon Dharma*, I referred to the Millennial generation as the Most Important Generation to ever Live on Earth (MIGLE) as they have to lead this human evolution towards the Butterfly phase[47]. Members of this generation are the "Miglets." I prefer this term to the nondescript "Millennials."

It is truly thrilling to see this amazing generation of Miglets lead our human metamorphosis into the Butterfly phase!

# 5. What is Our Relationship with the World?

*"You cannot get through a single day without having an impact on the world around you. What you do makes a difference. And you have to decide what kind of difference you want to make"* - Jane Goodall.

We like to think of ourselves as dominating the Earth. The factual evidence for this perspective is abundant. We are destroying forests, killing wild animals, birds and fishes with little to no opposition from them. We are raising and killing billions of domesticated animals as if they were mere widgets. We are poisoning the rivers and streams and installing treatment plants to produce potable fresh water for our own use, while condemning wild animals to drink poisoned water and fend for themselves.

Surely, we are dominating the planet, though we're not being particularly nice to other species. This is the standard perspective that mainstream environmentalists have adopted.

However in the process of "dominating" the Earth, we seem to have developed and deployed technologies that can continuously monitor the health of the planet, with sensors spanning the globe measuring every imaginable aspect of its proper functioning. We have developed and deployed technologies that scan space to detect every single asteroid or comet that can cause damage to the Earth. We are busy working on technologies that can defend the Earth and all Life on it from any catastrophes that can possibly occur.

That sounds like the kind of tasks that a true *"Khalifah"* or caregiver species would be doing, which is precisely what the Holy Quran and some interpretations of the Holy Bible had prescribed. But we seem to have accomplished these tasks almost unconsciously, as if compelled by a higher power. Therefore, what if our relationship with the world is truly like that of puppets to a master?

While such a relationship story is not very flattering for a proud species to contemplate, science is increasingly corroborating this scenario, just as it was envisioned in most religious scriptures. After all, would an omnipresent, omniscient, omnipotent deity allow a mere mortal species the free will to destroy Creation?

In the Hindu tradition I grew up in, it was commonly accepted that the breath was the "string" that connected each puppet to the Master. As long as we can take the next breath, we can be sure that the Master still needed the puppet for some larger purpose.

Therefore, what if science and religion can now agree that humans don't have free will? Imagine that we never had it and we never will. Wouldn't that make a truly revolutionary change in our relationship with the world at large? We would then stop blaming each other for the environmental and socioeconomic mess that we have made. We would be truly motivated to preserve and nurture the precious biodiversity that remains.

For if free will does not exist, then it would make sense to be routinely humble, grateful, compassionate and forgiving of oneself and fellow beings on Earth.

## 5.1 The Question of Free Will

The question of free will is an age-old one that has exercised minds for millennia. Currently, Neuroscience has advanced to a point where it is possible for a computer to accurately predict the choices that an individual is going to make, well before he or she is conscious of making those choices[1]. Therefore, many modern scientists and philosophers have concluded that we don't have free will. Says Daniel Do in his TED talk[2],

> "Free will cannot be grounded in logic, science or experience. We can logically deduce that it is incompatible with the laws of physics, experimentally verify that choice is the product of unconscious neurological processes, and observe through careful introspection that our sense of control is an illusion."

However, choices always stem from beliefs and beliefs can be reconfigured in our minds even if the resulting choices are not under our conscious control. This is a subtle distinction that was brought to light in the Hindu epic, Mahabharata, thousands of years ago.

In that epic, Lord Krishna was counseling the warring cousins, Arjuna and Duryodhana. But Duryodhana, who symbolizes material desire, did not wish for any advice from Lord Krishna! He merely stated,

"I know what's right, but I have no interest in doing it.

I know what's wrong, but I can't stop myself from doing it.

There is a strong tendency in me and I don't know how to stop it."

In contrast, Arjuna, who symbolizes human willpower, sought the counsel of Lord Krishna thus [3]:

"By what force does a person do wrong things? And how can I stop myself from doing them?"

In reply, Lord Krishna explained how anyone could reconfigure his or her mind to follow the path of Dharma. While the tendencies (*vasanas*) in our subconscious minds are uncontrollable and the thoughts (*vritti*) and actions (*karma*) that emanate from our conscious minds are also uncontrollable, the discriminatory capacity of our intellect is like a gatekeeper between the two and can be trained to make the right choices. Where there is a conflict between a choice that is good (*shreya*) and a choice that is pleasant (*preya*), we must align ourselves with the good choice and actively do something to eliminate the wrong, but pleasant choice. For example, if we now know that eating meat is disastrous for the planet and all Life on it, we can deem it immoral to consume meat since it deliberately harms other beings. Then, given a choice between plant-based foods (*shreya*) and animal-based foods (*preya*), we will begin to automatically choose the plant-based foods as we would have reconfigured our belief system and eliminated the wrong choice on ethical and moral grounds.

This is the purpose of meditation as well. In Vipassana meditation that the Buddha originally taught, we actively train our minds to become immune to likes and dislikes so that we can automatically begin to choose that which is right over a conflicting choice that is merely pleasant. For the happiness we experience from doing that which is in the greater good vastly exceeds the fleeting pleasure that we get from doing that which is merely self-indulgent. The founder of the modern discipline of positive psychology, Prof. Martin Seligman, has established that the happiness we experience in any action has three components:

Happiness = Pleasure + Engagement + Meaning.

That which is merely pleasurable gives us a certain amount of happiness. That which engages our skills further increases that level of happiness. That which has meaning for us, which is oriented towards a purpose larger than ourselves, vastly increases that level of happiness[4].

Universally, all of us are happiness seeking beings. This includes everyone from the President of the United States to a little mouse. Any crass selfishness that we exhibit is also aimed towards that pursuit of happiness. But Prof. Seligman has scientifically shown that selflessness leads to the greatest experiences of happiness, far greater than what can be obtained through crass selfishness. In controlled experiments, Prof. Seligman showed that subjects remember a day they spent serving the homeless in a soup kitchen far more fondly and clearly than the day they spent at a movie theater watching a good movie[5].

Therefore, selflessness is the highest form of selfishness! But the majority of human beings have not caught on to this simple, scientific fact. Perhaps the all-pervasive advertising in our Caterpillar culture has something to do with that confusion. Manufacturers routinely tout the consumption of their products as the very epitome of happiness!

From a biological standpoint, we don't have free will because what we do is strictly determined by the neurological circuitry in our brains. Knowing this, we can stop blaming people for their selfish or boorish actions and start helping them with their rehabilitation. For we also know that we have the capacity to rewire the neurological circuitry in our brains, if we so choose. We can exercise our powers of mental concentration and discrimination to overcome any negative tendencies so that we truly become an instrument of good. This is why Swami Vivekananda had stated that the purpose of education must be to teach mental concentration[6]. But this is difficult to achieve when we are disempowered in a socioeconomic system that is fundamentally rooted in fear.

Since time immemorial, Hindu philosophers have asserted that free will is an illusion. The only freedom we have is the freedom to let go. To let go of our ego and to let God work through us, just as our minds work through the cells in our physical body. Swami Vivekananda's guru, Sri Ramakrishna Paramahansa said[7],

> "As long as a man has not realized God, he thinks he is free. It is God Himself who keeps this error in man. Otherwise sin would have multiplied. Man would not have been afraid of sin, and there would have been no punishment for it. But do you know the attitude of one who has realized God? He feels: 'I am the machine, and Thou, O Lord, art the Operator. I am the house and Thou art the Indweller. I am the chariot and Thou art the Driver. I move as Thou movest me; I speak as Thou makest me speak.'"

This might seem paradoxical, coming from one of the most influential 19th century exponents of Advaita Vedanta, a philosophy based on the creation story of Mystical Unity. For how could it be possible that you are one with God, and simultaneously, have no free will? Shouldn't a God-realized human be able to do anything?

But one who has realized God deifies the world and sees everything as sacred. Such a person sees God in every being, to be loved unconditionally. Therefore, if the way of love is always your way, then what free will is there really in your actions?

Imagine a shopper in a supermarket filled with goods and produce. The shopper has many choices, rows and rows of processed foods as well as vegetables and fruits in the produce section. As long as marketing and advertising forces influence the shopper, he thinks he has the free will to choose what he wants. He wanders among the processed food aisles in the center of the supermarket and buys foodstuffs that are probably going to make him sick. But an enlightened shopper would know precisely what she needs and make a beeline, most likely for the produce section at the edge of the supermarket.

The Neuroscientific discovery of our lack of free will has the potential to unite us all towards advocating compassion for all Creation in our relationship with the world, since every religion in the world is also advocating the same compassion. In fact, in this truly watershed moment in human history, the notable atheist, Sam Harris, is now advocating for compassion just as loudly as Pope Francis[8]! That is true unity, indeed!

## 5.2 The Evolution of Needs

Any relationship is a connection established to meet the needs of the two parties involved. Our relationship with the Earth is no different. Ideally, our relationship with the Earth must meet our needs as a species while also meeting the needs of the Earth and other beings on Earth, each giving to the other in a way that creates a mutually beneficial connection. In a slight variation on the standard Maslow's hierarchy of needs, the motivational speaker, Tony Robbins, breaks down human needs as follows[9]:

1. Need for Security/Certainty
2. Need for Novelty/Uncertainty
3. Need for Significance
4. Need for Connection/Love
5. Need for Growth
6. Need for Contribution/Giving

The first four needs are what he calls as "personality needs," which are necessarily met in some fashion or the other. The last two needs are spiritual needs, which are not always met, but when met, lead to true fulfillment and enhanced happiness.

During the Caterpillar phase of our existence, the human relationship with the world, born of fear, is based on domination and violence. In the early days of our Caterpillar phase, the core Dharma of our species, compassion for all Creation or kindness to all Life, had to be routinely violated in order for our ancestors to survive. After all, a kindly, peaceful presence is not an appropriate response when confronted by a hungry, saber-toothed tiger. Therefore, our ancestors were forced to teach their children to harden themselves, as a survival skill. They developed weapons to protect themselves and eventually turned the tables around and began to dominate over all other creatures. Scientists have discovered that the most reliable indicator of human arrival in any land mass on Earth, whether in Asia, Europe, Australia or North America, is the disappearance of large megafauna from that land mass. Evidently, the first thing that our human ancestors did when they arrived anywhere was to kill all the large megafauna that might otherwise have killed them instead[10].

During the Caterpillar phase, the domination paradigm meets the security needs of human beings through the exercise of violence and raw power.

The need for novelty is met through the process of material discovery and progress in the pursuit of a comfortable environment.

The violence that comes with domination meets the need for significance as other beings surely tremble before us.

The need for connection with the Earth is met through our pets and our agricultural endeavors.

But during the Caterpillar phase, it is very difficult for human beings to meet the spiritual needs, the need for personal growth and the need to contribute beyond oneself towards the Earth. We tend to substitute personal growth with physical and material growth through mindless consumption, which is spiritually unfulfilling. Such consumption makes it very difficult for us to meet our need to give freely towards a cause larger than ourselves.

Once people realized the difficulty in meeting their spiritual needs in the Caterpillar phase, they attempted to transition into the Butterfly phase. Pockets of steady-state human civilizations developed where people learned to coexist relatively harmoniously among themselves and with other beings in their environment. The purpose of these civilizations was to help people fulfill their spiritual needs in addition to their personality needs. These steady-state civilizations weren't entirely nonviolent toward other animals and were therefore, not quite in the Butterfly phase.

In such civilizations, it is the community that meets the security needs of human beings.

The need for novelty is met through the arts or the search for enlightenment.

The need for significance is met through the achievement of excellence in some art or craft.

The need for connection is met through love and physical contact with the Earth.

The need for growth is met in the mental realm or the spiritual realm as opposed to the physical realm or the material realm.

The need to give freely of oneself is met in the service of the Earth community.

Such steady-state civilizations developed even as early as 60,000 years ago in places such as Australia[11]. In many cases, steady-state civilizations developed only after local climactic changes, desertification and other environmental catastrophes awakened people to the consequences of continuing with their domination paradigm in the Caterpillar phase. Steady-state civilizations existed in the Iroquois Confederacy and other indigenous communities of North America during the 12-15th centuries and still exist in indigenous communities throughout the world where people are generally contented and strive to meet their spiritual needs[12]. Historically, steady-state civilizations were not confined to small groups of people either. In his 1909 book, *Hind Swaraj*[13], Mahatma Gandhi wrote about the Indian civilization of the early 20th century as one such civilization where over 270 million people were generally contented with their lot, without constantly trying to "improve" some material aspect of their lives. This pervasive attitude of the Indians used to infuriate the British colonial rulers who ascribed it to "laziness," but it was really due to the Indian's perceived lack of necessity for such material growth. Gandhi used the example of the plough whose design had been unchanged for over 1000 years in India, not because Indians were incapable of improving the design, but because Indians didn't feel the need for such improvement. In contrast, he pointed out that people in the Western civilization were still relentlessly improving every material product in a continual, competitive quest for greater profits. Over a thousand years ago, India had already imbued in her children the idea that spiritual growth was far more consequential and rewarding than material growth and bodily comforts, which only served to atrophy the physical health of human beings. Though these early steady-state civilizations were not utopian for all, they were successful at reaching equilibrium in Nature as hundreds of millions of people were living in an India with 90% forest cover, as late as the 18th century[14].

But such stable, steady-state civilizations also let their weapons technologies stagnate and thus became vulnerable to outsiders who were still in the Caterpillar phase and were continuing to develop weapons, seeking to dominate not just the environment, but other humans as well. Slowly but surely, a global industrial civilization linked most such pockets of human communities together in a connected presence around the world. This is a civilization that originated in the West, grounded in science and characterized by its development and use of sophisticated tools and technologies to meet its needs and desires. In the Caterpillar phase, this global industrial civilization is based on a hierarchical social structure derived from the domination paradigm and is rife with inequities and oppressions, as a consequence. Now, we are getting feedback from a personal health standpoint, from a compassionate standpoint and from an environmental standpoint, that it is time to evolve to an appropriate steady state version within this global industrial civilization. The loudest feedback of all is now coming from an environmental standpoint, through imbalances in various bio-geophysical life support systems on Earth. It is clear that the present human relationship with the Earth, based on the domination of Nature, cannot be sustained for long.

Though modern humans have been in existence for just 200,000 years or so, the ancestral split with our nearest living cousins, the bonobos and the chimpanzees, occurred about 3 million years ago. Those 3 million years have been truly unique in the Earth's history as this is when the Earth began transitioning between the cold ice ages and the warm interglacial periods in so-called Milankovitch cycles, with a periodicity of about 100,000 years[15]. The transitions were not smooth since the Earth's climate is a highly nonlinear system, exhibiting relatively rapid phase transitions from one metastable state to another. Global surface temperature changes on the order of 6°C (10°F) were common during those transitions, but over a period of thousands of years. During all those climactic gyrations, human impact on the planet was minimal at best, though all that changed with the advent of agriculture about 10,000 years ago.

### 5.3 The Early Anthropocene Theory (EAT)

The Earth transitioned into the Holocene era about 12,000 years ago, an interglacial period of remarkable climactic stability, which allowed organized human civilizations to flourish[16]. Environmentalists and climate scientists have mainly attributed this remarkable climactic stability to good fortune. Then, they typically lament that we have squandered this good fortune in the recent past. For instance, Bill McKibben wrote in his book, *Eaarth*[17],

"For the ten thousand years that constitute human civilization, we've existed in the sweetest of sweet spots. The temperature has barely budged; globally averaged, it's swung between 58 and 60 degrees Fahrenheit. That's warm enough that the ice sheets retreated from the centers of our continents so we could grow grain, but cold enough that mountain glaciers provided drinking and irrigation water to those plains and valleys year round... We have built our great cities next to seas that have remained tame and level or at altitudes high enough that disease bearing mosquitoes could not overwinter...

But we no longer live on that planet. In the four decades since, that earth has changed in profound ways ... We're every year less the oasis and more the desert. The world hasn't ended, but the world as we know it has—even if we don't quite know it yet. We imagine we still live back on that old planet, that the disturbances we see around us are the old random and freakish kind. But they're not. It's a different place. A different planet. It needs a new name. Eaarth."

McKibben dates our intransigence as beginning in the 1970s, when ecologists such as Paul Ehrlich warned us all about human overpopulation and political leaders such as Prime Minister Indira Gandhi of India attempted to forcibly sterilize people. Some others have timed our intransigence at the start of the industrial era, about 200 years ago, when we began using fossil fuels in earnest.

Enter Dr. Bill Ruddiman, a Paleoclimatologist and Professor Emeritus at the University of Virginia. In 2003, he published the Early Anthropocene Hypothesis (EAH), claiming that the remarkable climactic stability in the Holocene era was not natural, but due to early human interference in the Earth's climate. He pointed out that we have been deforesting and denuding land for thousands of years and the greenhouse gas emissions from such agricultural land use changes would have prevented the Earth from retreating into another ice age. It is no coincidence that the vast contiguous desert in the mid-equatorial latitudes of the Earth today is where most well known ancient civilizations of the world arose.

The desert that begins with the Sahara at the western edge of North Africa extends all the way into India as the Thar desert of Rajasthan and into China as the Gobi desert. This vast contiguous expanse is the result of desertification caused by the Egyptian civilization, the Sumerian civilization, the Babylonian civilization, the Persian civilization, the Indus Valley civilization and the Chinese civilization, combined. Even today, we can witness the desertification occurring on the ground primarily due to human activities, in places like Rajasthan, India. Therefore, human beings have been impacting the environment tremendously since the agricultural revolution began. In fact, Prof. Ruddiman estimates that the net greenhouse gas emissions from land use changes over the past 10,000 years exceeds the net greenhouse gas emissions from all fossil fuel combustion in the industrial era, combined!

EAH is a hypothesis that dramatically alters the prevailing narrative and therefore, it has received considerable criticisms. However, all the scientific criticisms of the EAH published between 2004 and 2010 have been refuted by subsequent work. In his recent plenary lecture at the annual meeting of the American Association of Geographers, Prof. Ruddiman presented a blow-by-blow account of the main criticisms and their refutations[18]:

> "A model simulation of global methane ($CH_4$) sources claimed that anthropogenic emissions were not needed to explain the $CH_4$ increase after 5000 years ago, but archaeobotanist Dorian Fuller and colleagues showed that the spread of irrigated rice explains 70% of the $CH_4$ increase, and livestock will likely explain much of the rest.
>
> Initial simulations by land-use modelers suggested minimal pre-industrial carbon emissions, but landscape ecologist Erle Ellis, ecological modeler Jed Kaplan and I found historical records of larger per-capita land-use in pre-industrial time. Kaplan and colleagues published a historically validated land-use simulation indicating large Carbon DiOxide ($CO_2$) emissions closer to those in the EAH. Geographer Ralph Fyfe

and colleagues showed deforestation of Europe before industrial times based on pollen.

Geochemists claimed that changes in the $d^{13}C$ composition of $CO2$ rules out large pre-industrial carbon emissions. But large terrestrial carbon burial in boreal peats compiled by paleoecologist Zicheng Yu invalidated their mass-balance analysis.

Geochemists claimed that the closest interglacial analog to the Holocene (stage 11) disproved the EAH, but palynologist Chronis Tzedakis and colleagues showed that stage 11 is not a valid analog and that the best analog (stage 19) shows a $CO2$ decrease like that predicted in the EAH."

It is now safe to say that after thirteen years of corroboration, the Early Anthropocene Hypothesis has earned the right to be treated as a theory, the Early Anthropocene Theory (EAT). Therefore, far from being fortuitous, it is almost certain now that our human ancestors had been actively involved in maintaining the remarkable climactic stability over the past 10,000 years, though unwittingly. In fact, through the ice core records, we can even pinpoint the time when human impact on the Earth's atmospheric greenhouse gas composition began to show its signature in the atmospheric composition of CH4 and CO2. It began to occur around 5000-7000 years ago, and in that sense, the world as we know it today was indeed "created" at that time!

However, while our ancestors were unknowingly contributing to the stability of the Earth's climate in the Holocene era, it is only over the past 200 years, in the industrial era, that the human impact on the environment has reached such epic proportions that the Earth's bio-geophysical processes are becoming precariously unbalanced. It is now the scientific consensus that the systems of human civilization will need to come into conscious alignment with the Earth's planetary life support systems and play an integral part in the Earth's evolution into the future.

## 5.4 Human Impact in the Industrial Era

The industrial era is when most of the technologies that we see today got invented. In the process of creating and deploying these technologies, we have had to burn a lot of fossil fuels, destroy a lot of forests, pour a lot of fertilizers and irrigate a lot of land. Climate change is a consequence of all that activity. During the ice ages, the global surface temperature changes were caused by slight wobbles in the Earth's orbit around the sun. These wobbles result in slightly different amounts of solar energy reaching the Earth as the orbit changes in the Milankovitch cycles. When expressed in terms of "radiative forcing," which is the average solar energy that is absorbed per square meter of the Earth's surface per second, the variation due to Milankovitch cycles is on the order of 0.5 Watts/m$^2$. That may not seem like much, just 0.5% of the energy expended in a 100-Watt light bulb. When integrated over the entire surface of the Earth and over thousands of years, it is the difference between hot summer days versus a mile of ice on top of Chicago! In contrast, the greenhouse gases that humans have emitted to date are trapping an extra 3 Watts/m2 of solar energy, which is six times larger than the forcing due to the Milankovitch cycles[19]! The Earth's climate system is now beginning to react to this human induced forcing. As a result, the rate of increase of global surface temperatures today is also an order of magnitude faster than those due to the Milankovitch cycles.

If global surface temperature changes happen slowly, or for that matter, if changes in any of the planet's other environmental characteristics happen slowly, the Earth's biosphere is able to gradually evolve and adapt to them. As part of the Earth's natural healing process, the fauna migrate to adapt to the changes and then the flora follow more slowly as the old forests give way to the new. But if surface temperature changes occur quickly as they are doing now, it's as if the Earth has developed a fever since the biosphere hasn't had time to adapt. At present, the average global surface temperature on Earth has increased by about 1.0$^O$C (1.8$^O$F) since pre-industrial times, which is as if the Earth has developed a 38$^O$C (100$^O$F) fever, in human terms. Now, a 38$^O$C (100$^O$F) fever is not too debilitating, though the patient could probably use an aspirin to mitigate the fever. But if the fever increases to 39$^{\circ}$C (102$^{\circ}$F) or higher, then more drastic measures would be called for. Similarly, climate scientists have estimated that if the Earth's average global surface temperature increases more than 2$^{\circ}$C (3.6$^{\circ}$F) from pre-industrial levels, then other climate feedback loops would kick in causing the surface temperature of the Earth to spiral out of control[20]. Indeed, these days, even the 2$^{\circ}$C increase is being seen as too dangerous and perhaps we must limit the increase to less than 1.5$^{\circ}$C. But our current and future actions will decide whether the Earth's fever reaches such dangerous levels, for this fever is primarily caused by human activities on Earth.

But there's more to this story than just the fever. If the patient also has a coconut-sized growth by the side of his head, then a competent doctor ought to be treating that growth as well. Scientists at the Global Footprint Network have distilled human impact into a single figure, the ecological footprint, which they then compare with the available biological capacity of the planet[21]. According to their calculations, the human ecological footprint has been exceeding the available biological capacity of the Earth since the early 1970s and it is currently in excess of that capacity by nearly 60%. In effect, we have been eating into the ecological capital of the planet for the past four decades. Clearly, this is unsustainable, meaning that this imbalance will end at some point in time. We can let this rebalancing happen abruptly as the Earth's bio-geophysical systems break down catastrophically in response to our continued demands on the Earth and collapse our industrial civilization. Or we can voluntarily reduce our per capita ecological footprint and constrain it to be lower than the biological capacity of the planet as we transition towards a steady-state mode of living.

Again, our future actions will decide how the Earth evolves.

Systems scientists at MIT created elaborate models in the 1970s to study the interplay between human economic systems and the planet's bio-geophysical systems and thereby predict when the current trajectory of the global industrial civilization reaches a breaking point, before which the transition to a steady-state civilization must ideally happen[22]. Their models tracked six variables:

1) non-renewable resources remaining;
2) food per capita;
3) services per capita;
4) industrial output per capita;
5) global pollution; and
6) human population.

When they published the results of their simulations, including the predicted trajectories of these variables, in the book, *Limits to Growth*, in 1972, it became the best selling environmental book of all time, selling over 20 million copies worldwide! Their simulations predicted that human systems will reach a breaking point around 2025 in our present course leading to a civilizational collapse, but their models and simulated results were not taken seriously by the economists of that era [23]. But in 2010, Dr. Graham Turner, a scientist working at the University of Melbourne in Australia, compared the 1972 predictions in *Limits to Growth* with the actual data from 1970 through 2000 and found an almost exact match between them for all six of the variables[24]. Therefore, it is likely that a massive transformation of the global industrial civilization to a steady state mode of living will need to occur within the next decade or so before the predicted system collapse.

In the MIT systems model, the predicted breakdown in human economic systems occurs because the Earth's remaining non-renewable resources gets depleted over time. But that's a simplified, single variable stand-in for all of the Earth's complex bio-geophysical processes. Lately, in 2009, a group of 28 leading Earth systems and environmental scientists from different fields of expertise gathered to consider the various planetary bio-geophysical cycles and assess which ones are being stressed by human systems beyond tolerable limits[25]. They proposed a framework of planetary boundaries defining a safe operating space for human systems to work within. Of the nine planetary life support systems that they considered, they quantified limits on seven of them and estimated that three of the boundaries have been crossed already. Here are the nine Earth's life support systems and the scientists' findings on them, ranked in the order of the most violated limit to the least:

1. **Loss of Species:** Currently, upwards of 100 species are going extinct per million species extant each year. The safe operating limit for this variable is estimated to be 10 species/million/year and therefore, human systems are causing tenfold as much extinction of species as is safely tolerable. However, some biologists have claimed that even the proposed extinction safe limit of 10 species/million/year is far too negligent since the background rate is less than 0.1 species/million/year. But suffice it to say that the current rate of extinction is orders of magnitude above acceptable limits for us to sustain.

2. **BioGeoChemical Processes:** Humans are extracting over 120 million tons of Nitrogen from the atmosphere each year, thereby interfering with the planetary nitrogen cycle. The scientists estimated that the safe limit is around 35 million tons of Nitrogen extracted each year, which means that the human extraction is 3.5 times the safe limit. Of the other biogeochemical process, the anthropogenic phosphorous that is being pumped into the ocean, the scientists found that the human contribution of 9 million tons per year is still below the safe limit of 11 million tons per year. Prior to the industrial revolution, human activities had no impact on both these biogeochemical processes.

3. **Climate Change:** The limit for climate change was estimated in two ways, a) by using the atmospheric $CO_2$ (Carbon Dioxide) concentration and b) by using the anthropogenic radiative forcing. In terms of $CO_2$ concentration, the scientists reiterated Dr. Jim Hansen's famous limit of 350 ppm (parts per million)[26], which is the limit that spawned Bill McKibben's global environmental organization, 350.org. At present, atmospheric $CO_2$ concentrations are over 400 ppm, which is higher than the limit and violates it by about 50ppm. With respect to pre-industrial atmospheric $CO_2$ levels of 280ppm, the current level is 1.7 times the allowed safe limit of excursion.

This limit is expressed in terms of just the atmospheric CO2 limit, even though human systems emit a lot of different greenhouse gases such as methane, nitrous oxide, black carbon, etc., and not just CO2. However, human systems also emit a number of aerosols, which function to cancel the effect of greenhouse gases and cool the Earth's surface. It is just an accident of our global emissions profile that the net effect of the other short-lived greenhouse gases such as methane, nitrous oxide, etc., are roughly equal and opposite to the effect of all the aerosols that we emit, mostly as a byproduct of our fossil fuel burning. Therefore, the atmospheric CO2 limit by itself serves as a proxy for the overall contribution of human activities, for the time being.

Scientists estimated the alternative radiative forcing limit to be 1.0 Watt/m$^2$. While greenhouse gases emitted through human activities contribute 3.0 Watts/m$^2$ of radiative forcing, it is estimated that the aerosols negate about 1.5 Watts/m$^2$ of that forcing. Therefore, the net human induced radiative forcing is about 1.5 Watts/m$^2$ which violates the prescribed limit of 1.0 Watt/m$^2$ by 50%.

4. **Ocean Acidification:** Almost a quarter of the anthropogenic CO2 emissions is being absorbed in the ocean where the CO2 combines with water to become carbonic acid. As a result, the overall acidity of the ocean increases and this especially affects mollusks, clams and other shelled creatures that find it more difficult to form their shells. However, the scientists estimated that the safe limit for ocean acidity has not yet been breached though the increase in acidity since pre-industrial times is at 80% of the safe limit.

5. **Land Use:** The scientists estimated that the safe percentage limit of the Earth's land that can be converted to cropland is 15%, while the actual conversion that has occurred is 75% of the safe limit.

6. **Fresh Water use:** The scientists estimated that the safe limit for the total amount of fresh water that can be used by humans is 4000 km3/yr, while the current usage is 62% of the safe limit.

7. **Ozone Depletion:** The scientists estimated that the safe limit for ozone depletion has not yet been breached and the human activities have resulted in ozone depletion that is at 50% of the safe limit.

8. **Atmospheric Aerosol Loading:** Currently, about half of the radiative forcing due to anthropogenic greenhouse gases is being masked by the human emissions of aerosols such as SO2 (sulphur dioxide). However, aerosols cause respiratory problems in humans and other animals, and are estimated to be responsible for about 800,000 premature human deaths annually[27]. Aerosols also cause acid rain, affect the monsoons and global circulation systems. However, the scientists could not estimate a safe limit for the human induced aerosol loading of the atmosphere.

9. **Chemical Pollution:** The safe limits on the chemical pollution of the Earth's land, air and sea due to human industrial activities have not yet been quantified. Chemical compounds such as insecticides, pesticides, PCBs, dioxins and other so-called Persistent Organic Pollutants (POPs) bioaccumulate in living beings leading to various disorders and cancers. Other pollutants such as herbicides, excreted pharmaceuticals, heavy metals and radionuclides have potentially irreversible and harmful effects on many biological organisms, including humans.

### 5.5. The Earth Doctor's Diagnosis

Imagine that we are Earth doctors and we are being asked to diagnose the Earth's condition and prescribe a course of action for healing and reversal. Let's take each of the planetary boundaries and examine the root cause for the human induced perturbations in the underlying Earth processes.

Take the number one violated limit in the scientists' list: the unsustainable extinction of species on the planet. Species are stressed by multiple factors including habitat loss, invasive species, pollution, climate change, and just plain human consumption. But habitat loss through deforestation, desertification and egregious fishing practices such as bottom trawling and drag-net fishing, has been estimated to be responsible for 80% of species extinctions[28].

As for habitat loss on land, in the latest UN Intergovernmental Panel on Climate Change (IPCC) Fifth Assessment Report (AR5), Working Group 3, Chapter 11 breaks down the land use and human biomass consumption figures, sector by sector, from the year 2000, as follows[29].

72% of the useable landmass of the Earth has been appropriated for human purposes, while 8% is untouched, pristine forest and 20% is abandoned, unused or naturally regenerating land.

11.77 Giga tons of dry matter plant biomass is extracted from all that land for human use. Of that, 60% goes to feed our livestock, 15% is used for bio-fuels, mainly as fuelwood, 13% is used by the processed food industry, 10% goes towards other industrial uses and only 2% is consumed directly by humans in the form of fruits, vegetables, nuts, seeds, grains, etc.

Of the 7.01 Gig tons of plant biomass fed to livestock, we end up with only 0.18 Giga tons of meat, dairy and eggs on our dinner plates, an enormous 39-fold reduction in biomass due to trophic losses and waste.

In contrast, of the 1.78 Giga tons of plant biomass that we extract for human consumption in the form of plant foods, 1.36 Giga tons ends up on our dinner plates, a mere 25% reduction in biomass, mainly due to processing.

We know that the biomass of 7.4 billion humans alone (500 Million metric Tons (MT)) exceeds the biomass of all megafauna (200MT) that existed on Earth between 10K-100K years ago by a factor of 2.5[30]. But as far as the planet is concerned, we are presenting the profile of 44.4 billion humans, not just 7.4 billion humans, once we include the biomass consumed by our domesticated animals!

Since the year 2000, the impact of Animal Agriculture has only become more pronounced since the industry has grown by 30% in that time span. The International Livestock Research Institute (ILRI) scientists estimate that livestock systems occupy 45% of the land area of the planet today. The IPCC has estimated the same figure to be 40%, but as of the year 2000. Where livestock systems occupy land, biodiversity is greatly diminished - for example, wild animals are killed as "intruders" and wild plant species are replaced with fodder - and species extinctions inevitably occur. In the ocean, our fishing practices are so indiscriminately violent and egregious that, for example, we kill 6 times as much marine life as by-catch than the shrimp we harvest[31]. This is why the Center for Biological Diversity, a grassroots environmental organization that is committed to preventing species extinctions, has begun a "Take Extinction off Your Plate" campaign to promote the wide scale adoption of plant-based diets[32].

The second major safe limit violation, the Nitrogen cycle, is also primarily due to Animal Agriculture since more than half the industrially grown mono-cultured crops of the world that rely on nitrogen fertilizers for production are fed to animals directly.

With respect to the third safe limit violation, the Carbon cycle, human burning of fossil fuels is clearly the number one cause, responsible for an estimated 57% of all anthropogenic greenhouse gas emissions, using a 100-year window for calculating $CO_2$ equivalences of methane and other short-lived greenhouse gases. However, many scientists have questioned the use of a 100-year time window for calculating $CO_2$ equivalences of short-lived greenhouse gases, given the urgent nature of climate change. With a 20-year time window, for instance, methane becomes three times as powerful a greenhouse gas, than over a 100-year time window.

But even here, Animal Agriculture is a substantial contributor to anthropogenic greenhouse gas emissions, with end-to-end lifecycle estimates ranging as high as 51% of the total[33], if we use 20 year time windows for calculating $CO_2$ equivalents. A good chunk of the fossil fuels are burnt to support this industry for the transportation of feed to animals and the processing, refrigeration and transportation of animal carcasses. In addition, a major portion of the greenhouse gas contributions due to deforestation is because of Animal Agriculture, and a major portion of the soil carbon emissions due to desertification is also because of Animal Agriculture.

There is an ongoing debate on the magnitude of the role of Animal Agriculture in anthropogenic greenhouse gas emissions. While scientists are supposed to be unbiased observers in the scientific studies they are conducting, such biases are inevitable when personal habits are related to the studies. While estimating the lifecycle impact of Animal Agriculture, almost every scientist is prone to such bias, since the vast majority of scientists in the world consume animal products on a routine basis. For instance, at the largest annual gathering of climate scientists in the world, the American Geophysical Union (AGU) Fall Meetings in San Francisco where 25,000 scientists gathered in 2015, the banquet dinner consisted of steak as the main course. Hardly anyone at the banquet ordered the lactovegetarian meal option, and I was the only one who special-ordered a plant-based, vegan meal. I know this because I was waiting for them to boil some pasta and vegetables for my dinner while the steak knives were clanging all around me.

The traditional framing of climate change is that the human burning of fossil fuels primarily causes it. Deforestation, which is mainly to support animal agriculture, adds a much smaller component. Land and the ocean absorb 55% of those emissions while the other 45% accumulates in the atmosphere causing climate change. With such a framing, we feel disempowered because fossil fuels are such an integral part of our daily lives these days.

But that is the official story. When we study the carbon cycle in detail using the UN IPCC AR5, we discover that the true story is much more nuanced. Fossil fuel burning emits 7.6 Giga tons of Carbon (GtC) annually. Deforestation emits 1.5 GtC annually. But other human activity on land and in the ocean causes 29.3 GtC to be emitted annually through decaying crop biomass, through raising billions of farm animals, through firewood burning, etc. That is nearly 4 times as much carbon emissions as the fossil fuel component! But human activity also causes 34.1 GtC to be sequestered annually through the use of nitrogen fertilizers, CO2 fertilization effects, irrigation and other technologies. The net result is a sequestration of 4.8 GtC, which is about 55% of the emissions due to fossil fuels and deforestation.

This means that, in total, human activity is causing 38.4 GtC to be emitted annually and 34.1 GtC to be sequestered annually. This is where the enormous impact of our food and other consumer choices is hidden – in plain sight.

In 2006, the UN Food and Agricultural Organization (FAO) published the initial estimate of lifecycle impact of Animal Agriculture[34]. The FAO report estimated that the Animal Agriculture industry was responsible for 18% of anthropogenic greenhouse gas emissions, about 50% more than the entire transportation sector, which was estimated to be responsible for 13%. Then, in 2009, two Environmental Assessment specialists from the World Bank Group, Dr. Robert Goodland of the World Bank and Jeff Anhang of the International Finance Corporation (IFC), using the UN IPCC Fourth Assessment Report (AR4) lifecycle impact analysis guidelines, estimated that the Animal Agriculture industry was responsible for at least 51% of all anthropogenic greenhouse gas emissions, in total. Their estimate was a lot higher than the FAO estimate because they included all the Tier 3 contributions that were missing in the FAO estimate, as per IPCC guidelines[35]. They used a 20-year window for calculating $CO_2$ equivalences of short-lived greenhouse gases such as methane, instead of the 100-year window used in the FAO estimate. Clearly, since climate change is an urgent problem that requires immediate action within the next 1-2 decades, using a 20-year window for calculating the impact of methane emissions is actually more appropriate[36].

When the WorldWatch Institute initially published the Goodland-Anhang estimate, the estimate needed to be debated and corroborated in a scientific journal before gaining legitimacy. The 18% figure from the 2006 UN FAO report was compiled by scientists who were employed by the Animal Agriculture industry (Steinfeld, Herrero et al. consult for the International Livestock Research Institute (ILRI), a think tank for the industry) and therefore, it was prone to be biased. For instance, all of the criticisms of the FAO estimate in the Goodland-Anhang report were clearly valid. Naturally, the ILRI scientists would be inclined to downplay the environmental impact of their industry products, just as the Tobacco Institute scientists were inclined to downplay the health impacts of tobacco. On the Internet, there were a lot of ad hoc criticisms of both estimates with bloggers making arbitrary corrections to either estimate and I described the situation thus in the 2011 book, *Carbon Dharma: The Occupation of Butterflies*[37]:

"We got an inkling of the impact of animal agriculture when the UN published its Livestock and Climate Change report in 2006, where the livestock sector was calculated to be contributing 50% more greenhouse gas emissions (18%) than the entire transportation sector of the world (12%). Later, in a 2009 Worldwatch Institute report, two UN Environmental Assessment (EA) specialists, Robert Goodland and Jeff Anhang, pointed out that the 2006 UN report failed to take into account the carbon cycle imbalances caused by the conversion of forests to livestock pasture lands. They came up with an estimate that the livestock sector was responsible for 51% of world greenhouse gas emissions but their calculations based on the breathing contribution of livestock were not widely accepted. However, in 2010, Prof. Danny Harvey of the University of Toronto in Canada, building upon the thesis work of Stefan Wirsenius from the Goteborg University in Sweden from 2000, calculated that the average human being is consuming more energy in food than in fuel and shelter combined, when we take into account the embedded plant-based energy input to the animal agriculture systems. And the main reason is that animal agriculture is so inefficient that, on an average, it requires 100 Joules of embedded plant-based energy to produce less than 4 Joules worth of animal foods such as eggs, dairy and meat for human consumption."

However, in late 2011, the scientific community hashed it out in the peer-reviewed Animal Feed Science and Technology (AFST) journal. The principal authors of the 2006 FAO report, Herrero et al., wrote a paper rebutting Goodland and Anhang's criticisms of their work[38]. Goodland and Anhang strongly responded to that rebuttal and basically reiterated their original calculations in the same journal[39]. Herrero et al. then declined to continue the debate[40].

As per the scientific process, the calculations of Goodland and Anhang stand until there is another peer-reviewed scientific paper that updates their methodology and calculations - perhaps using IPCC AR5 lifecycle analysis guidelines with updated CO2 equivalence factors for methane etc. - and goes through a similar peer-reviewed, scientific scrutiny. Please note that in their Worldwatch report, Goodland and Anhang state that Animal Agriculture is responsible for **at least** 51% of the global greenhouse gas emissions, i.e., the 51% fraction is a lower bound!

The ILRI scientists later doubled down on their earlier mistakes and reported a smaller 14.5% figure in the UN FAO report of 2013, without any explanation and without addressing the debate points in their AFST journal exchange with Goodland and Anhang[41]! That made their earlier estimate from 2006 even less credible. But then, how can we seriously expect scientists with deep ties to the Animal Agriculture industry to provide credible estimates on the harmful impact of that industry? That's like expecting scientists from the Tobacco Institute to provide credible reports on the harmful impacts of smoking.

Perhaps the most controversial aspect of the Goodland-Anhang analysis has been their use of estimated CO2 emissions from livestock breathing, as a proxy for soil carbon loss due to Animal Agriculture. The typical argument that critics use is that everyone breathes and therefore, it is all part of the "natural" carbon cycle. But they don't seem to use the same argument for not counting the methane contribution of livestock. Since all ruminants emit methane, why should livestock methane emissions be treated as anthropogenic? Indeed, the IPCC routinely does count livestock methane emissions as anthropogenic!

Both livestock CO2 and methane emissions need to be counted since the biomass of livestock today is an entirely man-made, extraordinary and unprecedented burden on the ecosystems of the Earth. According to paleobiologist, Prof. Anthony Barnosky of UC

Berkeley, the biomass of all megafauna stayed fairly constant at around 200MT from 100,000 years ago until about 10,000 years ago, when humans overspread the Earth and caused the Quaternary extinction of large megafauna[42]. From 10,000 years ago until about 400 years ago, there weren't too many species extinctions and the global biomass of megafauna slowly recovered to 200MT, mainly as a result of increased livestock and human population.

However, since the industrial era, the biomass of megafauna exploded from 200MT to about 1500MT today, with humans accounting for 500MT and our domesticated animals accounting for most of the rest, with the biomass of all wild megafauna reduced to a mere 2-3% of the total. The breathing contribution of all that excess biomass is clearly out of balance with the natural carbon cycle and it shows in the rapidly spreading deserts of the world. The breathing contribution of livestock stands out since our livestock population is an unnatural mix of mostly young animals that eat and metabolize at 2.5 times the rate as one would expect from their biomass alone. In fact, the IPCC AR5 figures show that our livestock population metabolizes 5 times as much dry matter biomass (4.69 Gt) as all human beings put together (0.93 Gt). Though our livestock weigh about 1000MT, they present the profile of a biomass that weighs 2500MT! All that excess metabolization results in $CO_2$ emissions and should be counted, even as per IPCC AR5 guidelines. Therefore, Goodand and Anhang's use of the breathing contribution of livestock as a proxy for soil carbon loss is appropriate and even has precedence in the scientific literature[43].

Fossil fuel burning produces significant $CO_2$ emissions, the prime culprit for ocean acidification. However, if the scope of Animal Agriculture diminishes considerably and regenerating forests sequester some of the excess $CO_2$ in the atmosphere, the acidity of the ocean will diminish as the dissolved $CO_2$ in the ocean outgases into the atmosphere again.

As for fresh water, 55% of human fresh water use is for Animal Agriculture in the US[44], while the fossil fuel industry accounts for about 25% of fresh water use[45].

With respect to ozone depletion, one of the primary mechanisms for it is during the conversion of methane to $CO_2$ in the upper atmosphere. In turn, Animal Agriculture is one of the primary sources of anthropogenic methane emissions[46]. However, among fossil fuels, natural gas systems leak methane as well, though the true magnitude of these leaks is in dispute.

With respect to aerosol loading, the major source of human emissions of aerosols is during the burning of fossil fuels[47].

Finally, with respect to chemical pollution, most of the herbicides, pesticides and insecticides are used in the agricultural production of animal feed. Most of the pharmaceuticals are ingested to treat human diseases that occur due to the consumption of animal foods. Therefore, if Animal Agriculture ceases, a significant portion of the chemical pollution on the planet will cease as well, though it will take a while for the existing pollution to be sequestered by regenerating forests[48].

As Earth physicians, let us imagine that we are asked to write a prescription for each and every human being to fill so that humanity's overall relationship with the Earth changes to a more benign, safer, steady state version. What should we tell people to do? Should we tell them to focus on weaning themselves from fossil fuel use, i.e., go solar, or should we encourage the ongoing transition away from animal products, i.e., go vegan?

The best way to answer this question is to do a sensitivity analysis through a simple thought experiment. Imagine that we wave a magic wand and get everybody to go vegan instantly. This immediately impacts 8 of the 9 planetary boundaries and instantly begins to reduce the violated safe limits as well. If everybody went vegan, it would immediately release 35-40% of the land area of the planet back to Nature, to regenerate forests and re-wild the planet[49].

1) Since habitat gets returned to wildlife, the safe limit violation on species extinction rates would be reduced immediately.

2) The safe limit violation on the nitrogen cycle gets reduced immediately as we replace the 0.18 Gt of dry biomass of animal-based foods, (and 0.09Gt of seafood biomass), with equivalent plant-based foods, derived from crops. This is because, as of 2000, according to the UN IPCC figures, 3.14 Gt of crop-based dry-matter biomass was going directly to feed livestock and almost all of that required nitrogen fertilizers[50]. Let's say we need to use twice as much plant crop biomass to produce the plant-based food equivalents for the 0.27Gt of animal-based foods. Then, we need to raise 0.54 Gt of plant crop foods for the plant-based substitutes instead of 3.14Gt, meaning that around 80% of those plant crops can be eliminated, thereby reducing nitrogen fertilizer use by 40% overall[51].

3) The safe limit violation on the carbon cycle gets reduced as regenerating forests sequester $CO_2$ from the atmosphere in trees, plants, insects, animals, birds, and in the soil. Along with Prof. Atul Jain of the University of Illinois at Urbana-Champaign and his graduate student, Shijie Shu, I presented a paper on this so-called "Lifestyle Carbon Dividend" at the AGU Fall meeting in December 2015[52]. Using the Integrated Science Assessment Model (ISAM) that Prof. Jain's team developed at the University of Illinois, we estimated that recovering forests can sequester 265 Giga tons of Carbon (GtC) on just the 41% of current grasslands that used to be forests in 1800. That is more carbon than has accumulated in the atmosphere since 1750 (240 GtC)!

4) As the $CO_2$ in the atmosphere gets sequestered in regenerating forests, the acidification of the ocean reduces as well, as the ocean outgases $CO_2$ to reach a new equilibrium.

5) Around 40% of the cropland can be returned to Nature thereby mitigating the land use planetary boundary.

6) Around 50% of the fresh water use is avoided as Animal Agriculture ceases. Besides regenerating forests create more fresh water as well.

7) As methane emissions are reduced, ozone depletion is ameliorated as well.

8) The regenerating forests begin to sequester chemical pollutants, and as humans stop ingesting chemical pollutants through animal-based foods and become healthier, their production of excreted pharmaceuticals also reduces.

Next, let us consider a second thought experiment. Imagine that we can wave the same magic wand and turn our entire energy infrastructure over to solar, wind and other clean sources instantly so that fossil fuels no longer have to be burnt to run the human enterprise. Let us assume that we leave everything else unchanged as at present so that we continue to eat animal foods and indeed, double our consumption by 2030 as envisioned by the UN FAO. As a result of this instant change to our energy infrastructure, we stop pumping a majority of the greenhouse gases, especially carbon dioxide ($CO_2$), into the atmosphere and we immediately stop exacerbating human impact on 4 of the 9 planetary boundaries. That is the good news. But the bad news is that,

1) The surface temperature of the planet will start increasing at a faster rate and it will increase approximately by an additional $0.5^{\circ}C$ ($1^{\circ}F$) within a decade as aerosols disappear from the atmosphere. This temperature increase occurs because when we burn fossil fuels, we not only pump greenhouse gases into the atmosphere, but we also simultaneously pump aerosols such as sulphur dioxide ($SO_2$) that are partially shielding us from the true impacts of our greenhouse gas emissions. While $CO_2$ persists for hundreds of years in the atmosphere, aerosols precipitate and disappear from the atmosphere within a couple of weeks thereby exposing us to the true effects of our greenhouse gas emissions to date. Think of aerosols as the haze that prevents the sun from shining brightly on our Earth. When the haze disappears, the Earth heats up.

Using the fever analogy, the Earth's temperature will rise from a $38^{\circ}C$ ($100^{\circ}F$) fever to a $38.5^{\circ}C$ ($101^{\circ}F$) fever within a decade even though we have stopped burning fossil fuels completely.

Personally, with a 100°F fever, I can still function in my day-to-day life, but with a 101°F fever, I'm resting in bed waiting for the fever to subside. That extra 1°F temperature increase can have serious repercussions for us on the planet as well. This is why the eminent climate scientist, Dr. Jim Hansen, has concluded that we need to keep pumping aerosols into the atmosphere even after we phase out fossil fuel burning in a Faustian bargain to avoid excessive temperature increases[53]. It is very likely we will have to do that by extracting the sulphur from coal in a chemical process and then burning the sulphur alone with little to no benefit whatsoever. And suffer the resulting consequences of acid rain falling on our heads.

2) The $CO_2$ in the atmosphere will not diminish for hundreds of years and therefore, the violated limit on the carbon cycle will stay violated for hundreds of years as well, even though we no longer burn any more fossil fuels. Furthermore, since our land use changes will continue unchecked, $CO_2$ emissions from deforestation, desertification and methane emissions will continue to add to the atmospheric $CO_2$ levels.

3) As the $CO_2$ concentration in the atmosphere increases due to deforestation, the ocean will continue to become more acidic, but not as much as it would with the additional burning of fossil fuels.

Therefore, in this second experiment, the conversion of the energy infrastructure does nothing to heal the climate and would actually make the situation worse.

As Earth doctors, our diagnosis task is a no-brainer, really. Considering all the symptoms that the Earth is exhibiting and considering the results of our thought experiment, we can now see that it is primarily Animal Agriculture, which has deliberate, institutionalized violence on other beings implemented on a gargantuan scale, that is the main source of the cancer underlying the Earth's fever. Just among land animals alone, 200 million of them are being directly slaughtered each and every day for food all over the world. If we count sea animals and the by-catch of sea animals during industrial fishing, the total is on the order of 1 billion animals each and every day to sustain the appetites of 7.4

billion human beings. Billions of animals are being routinely enslaved, forcibly impregnated, with their menstrual secretions harvested, their maternal secretions harvested, and then they are slaughtered, most when they are still just babies. Every one of these acts is an act of institutionalized violence and every form of animal husbandry on this planet performs almost every one of these acts of violence. Even the certified, highest rated, "humane" meat, dairy or egg producing small farm performs almost all these acts of violence on a routine basis. This is the primary reason for all the violated limits in the scientists' list. This is the "cancer" underlying the "fever," the cancer that is eating away at the Earth.

But this is like a rare form of cancer that is eminently curable. We are already waking up to reverse that cancer by voluntarily changing our consumption patterns. Despite the deafening silence on this issue from climate scientists and mainstream environmentalists, who are naturally frightened of the system implications as an industry directly and indirectly employing a billion people bites the dust, the Miglets are already leading a global transition towards veganism. Surely, as a species, we can figure out better ways to guarantee economic security for people without requiring them to destroy the planet?

However, that doesn't mean we should stop converting our energy infrastructure over to clean sources. In fact, the two steps, go vegan and go solar, will be accomplished simultaneously. Those of us in the global North who have access to food abundance will go vegan right away and then convert the energy infrastructure over to clean sources over the next 1-2 decades as solar technologies get produced in volume. This is the optimal order for the change to occur since the carbon sequestration due to the vegan transition will then compensate for the temperature increase due to the atmospheric aerosol reduction from the clean energy transition.

## 5.6. Modern Animal Husbandry

Modern animal husbandry is a marvel of technology. From instruments that guarantee the impregnation of animals to machines that milk dairy cows and the disassembly line that is the modern slaughterhouse, technology has relentlessly squeezed the resource requirements for livestock production. Writes James McWilliams[54],

"The efficiency of an industrial slaughterhouse, macabre as it may be, is a spectacle to behold. A farm animal entering the front door will reach the exit about 19 minutes later. It will do so not only as chops destined for the meat counter, but as pelts bound for Turkey, lungs sent to dog-treat manufacturers, bile for the pharmaceutical industry, caul fat (the lining of organs) for Native American communities, and liver destined for Saudi Arabia (which, go figure, distributes cow liver globally)."

But among all these technologies, the Confined Animal Feeding Operations (CAFOs) or "factory farms" stand out for their relentless efficiencies of production. Can you imagine being actually thankful for factory farms, those "gulags of despair"[55] where modern animal husbandry has been conducted in such deplorable conditions? Knowing that factory farms are universally unloved, some have claimed that if only animal husbandry could be conducted differently through rotational grazing or some such "improved intensification" technique, Animal Agriculture could become sustainable[56]. But the fact is that the mimicry of complex, bio-diverse ecosystems is difficult to achieve with just a few species: cows, humans, chickens and grass, especially when all these species are invasive in most places on Earth. When their biomass is constantly removed from land and consumed and excreted elsewhere, the nutrients on that land get depleted. Further, if animal husbandry is conducted within native bio-diverse ecosystems, then its impact on planetary boundaries would actually be far worse than with the factory farmed versions! This fact was vividly illustrated to me during a visit to the Zulu Nyala safari in South Africa, after attending the UN COP-17 Climate Change meeting in Durban in 2011.

The Zulu Nyala safari is situated on 5000 acres of lush green African land and it was completely encircled with an electric fence to create a somewhat closed ecosystem. On these 5000 acres lived over 1100 herbivores, including 380 impalas (deer), 250 Nyalas (deer), giraffes, wild boars, rhinos and elephants. In addition, there were exactly 7 carnivores, all of them cheetahs.

There were two cheetah brothers, a mother cheetah and her four 8-month old cubs. The two cheetah brothers were always found together, since one had a broken leg and couldn't hunt. Therefore, his brother used to hunt for the both of them. During all our Jeep treks, we would always find these cheetah brothers near the Eastern border of the safari because they were part of a litter of four and their two sisters were sold to the neighboring safari.

Our game warden, Jabulani Tembe, told us that the four cheetah cubs would be removed from the safari as soon as they reached their first birthday because the safari could not really support 7 cheetahs. This is because between them, these 7 cheetahs were consuming the equivalent of 1 whole deer each day. The Safari couldn't sustain the killing of 365 deer each year, since the deer do not reproduce fast enough to replenish such a kill rate.

Therefore, 5000 acres of lush green land in Africa could support over 1100 herbivores and exactly 3 cheetahs, on a sustained basis. Such is the arithmetic of land use requirement for supporting herbivores vs. carnivores in a truly bio-diverse, natural ecosystem.

If the cheetahs became human-like tinkerers of Nature and chose to engineer an ecosystem that maximized the production of food for them, they would

  1) Remove all other herbivores on the safari, leaving only deer, since the cheetahs don't or can't eat the other animals,

  2) Cull all deer as soon as their initial growth spurt is over so that the efficiency of conversion from plant foods to deer flesh is maximized,

  3) Keep the deer imprisoned and stationary in order to minimize the food energy that the deer expend in locomotion and maximize the food energy converted to muscle mass,

4) Institute a fossil-fuel based mechanical system to transport the plant food to the enslaved, stationary deer,

5) Institute a controlled impregnation program on the female deer population,

6) Institute a recycling program to process all left over deer carcasses into protein meal for the live deer and finally,

7) Inject the deer with antibiotics and hormones to enhance their growth of muscle mass.

The result would resemble the factory farms of modern animal husbandry. If we reverse the efficiency gains achieved in each of these seven steps and revert to "humane animal husbandry," wouldn't the land use requirement for the Animal Agriculture industry actually increase? Indeed, if we consider the enormous footprint of cheetahs in the wild, one cheetah supported on every 1700 acres, it is truly a prodigious engineering achievement on part of the Animal Agriculture industry to churn out so much animal-based foods for their billions of customers using just 45% of the land area of the planet! This works out to an average of just over 2 acres per person for a population of 7.4 billion people, 850 times less than what is required for cheetahs in the wild at the Zulu Nyala Safari!

But of course, just the top one-third of humanity is doing most of that consumption of animal foods, while the bottom 1 billion people are literally starving. Cheetahs are 100% obligate carnivores, unlike humans. Nevertheless, it is hard not to admire the accomplishments of the food scientists in the Animal Agriculture industry from an engineering standpoint. If not for all the hormones and the antibiotics to promote rapid muscle growth, the genetic selection that singled out the fastest growing animals and other ingenious technologies that these food scientists have deployed, all the forests in the world would probably have been completely destroyed by now. In our Butterfly phase, we would have had a much more difficult, if not impossible, healing job to do.

It is due to the enormous production efficiencies built into modern animal husbandry that it is still possible to restore much of the biodiversity of the planet, as was accomplished at the SAI sanctuary. Recently, scientists from the University of Arkansas[57] pointed out that if human babies grew at the exact same rate as broiler chickens in the poultry industry do today, a 2 month-old human baby would routinely weigh 300kgs or 660 lbs! But such a broiler chicken is growing at least six times more rapidly than a chicken in Nature, aided by the hormones, antibiotics and other chemical stimulants in the feed.

If we revert back to the bucolic ways of old, surely we would need 6 times the resources to raise the same biomass of chicken for human consumption?

## 5.7. Taking Off Our Blinders

It's a human condition to be blind to our foibles. It is very difficult for us to correct for that blindness when the dominant culture encourages and subsidizes those foibles.

Imagine a society where drugs are legal. Drug dealers are everywhere. They bombard people with ads. They target children. They wear clown costumes to entice children to sample their wares. They build play places in their drug dens to lure the children. They tell them that the drugs are good for them. They claim the drugs are the only source of nutrients for their bodies. The government supports the drug dealers with taxpayer-funded subsidies. It even runs ads for them for free.

Then addiction to that drug would be common in such a society. Scientists in that society would endorse that drug when they are themselves consuming it thrice a day, every day, even if that drug is responsible for 80% of the health care costs in that society. Just as the "doctors smoking Camels" endorsed cigarette smoking in the 1950s!

But the facts are undeniable. Animal agriculture, which is the primary avenue for institutionalized human violence towards non human beings, is also the primary reason for the environmental degradation on our planet. The feedback signals from Nature are getting louder and louder, telling us that it is time to change. We're getting sicker and sicker as we consume animal foods. We can't stand the animal cruelty that we're routinely seeing on the Internet. Yes, we try to avoid watching these videos, but we know they are there because some of our friends are talking about them and have shared them in our social networks. Super storms and mega droughts and year-round wildfires are buffeting us as the Earth indicates that we've gone too far with the environmental degradations. But just as the Second Hand Smoking campaign finally corralled the tobacco industry and brought it to heel, the notion of "Second Hand Eating" that the documentary, *Cowspiracy:The Sustainability Secret*, is popularizing, has the potential to finally corral the Animal Agriculture industry and render it impotent[58].

When I first came to the US as a graduate student in 1981, I was a smoker, an addict to a terrible habit that I had picked up a couple of years earlier. I was astounded that I could smoke on buses, trains, planes, supermarkets, restrooms and indeed, almost anywhere in the US. When I got married in 1984, I was even smoking in the apartment that I shared with Jaine. The US was the "land of the free" and freedom was about doing whatever you want, wherever you want.

But of course, our personal freedom must not impinge on someone else's freedom to enjoy "life, liberty and the pursuit of happiness." That's how the "Second Hand Smoking" campaign began. Activists pointed out that smokers were forcing people around them to breathe in carcinogens, thereby causing the bystanders to get ill and even cancer. Jaine contracted a persistent cough and her doctor told her that my smoking was causing the problem. Therefore, I began stepping outside our apartment to smoke cigarettes. When we moved to New Jersey, I even had to step outside in the snow during winter in order to smoke.

Then we imposed heavy taxes on cigarettes. We cranked up the health and life insurance premiums for smokers and ratcheted up the pressure to quit. We built special places where we could smoke. It was usually a dark, stinky room, tucked away in a corner. Or we built outdoor spots far away from foot traffic. Finally, Brazil went ahead and banned smoking in all outdoor places throughout the entire country!

The Tobacco industry is now on the run. It is trying to entice people in the global South to take up smoking. It uses ads with cartoon characters to entice children to take up smoking. But despite 51 years of relentless anti-smoking campaigning, the world tobacco consumption has yet to hit its peak after the campaign began in 1965.

It is hard to give up smoking. In my own case, it was my mother's demise which finally caused me to give it up, but in the process, I learnt a lesson that I'll never forget, that procrastination never pays. The Rev. Martin Luther King, Jr., put it succinctly[59],

"The time is always right to do the right thing."

But if we thought that "Second Hand Smoking" impinged on our right to lead a healthy life, wait till we all become truly aware of what "Second Hand Eating" is doing. The population of human beings alone is unsustainable on the planet since the biomass of our single species (500MT) exceeds the biomass of ALL wild megafauna from 10,000 years ago by a factor of 2.5! Someone eating a hamburger is causing the equivalent of an extra 2500MT of megafauna biomass, or 5 times our human biomass, to be grown on planet Earth. Each hamburger also requires 660 gallons of fresh water to produce[60]. As such, that hamburger consumer truly has the power of life or death over all Life, including the consumer himself.

Not only is our culture that condones the consumption of animal foods inexorably killing the consumer, it's inexorably suffocating all Life on Earth. While I had to leave my apartment and smoke cigarettes outside in order to avoid Second Hand Smoking affects, there is no special place to go on our planet to avoid Second Hand Eating effects. We, human beings, are social creatures and we certainly don't want to be seen as hurting our community, our friends, our neighbors, let alone our children, our grandchildren and the whole planet, in addition to our own selves. Therefore, despite the silence of climate scientists and environmentalists and the active discouragement of governments as they desperately try to preserve the current socioeconomic system, people are now transitioning away from animal-based foods in droves, the system notwithstanding!

When our choice affects the well being of billions of others around the world, it's no longer a personal choice, but a moral choice. When it comes to such a moral choice, Love is a much stronger force for social change than Fear.

## 5.8. Re-imagining our Relationship

The Nobel-winning behavioral economist, Daniel Kahnemann, said[61],

> "We can be blind to the obvious. And we can be blind to our blindness".

But the time is ripe for our globally connected industrial civilization to overcome culturally bequeathed habits that no longer serve their original purpose and transition to a truly nonviolent, sustainable, steady-state presence on Earth. As Albert Einstein said[62],

> "Our task must be to free ourselves by widening our circle of compassion to embrace all living creatures and the whole of nature and its beauty... Nothing will benefit human health and increase the chances for survival of life on earth as much as the evolution to a vegetarian diet."

The evolution to a vegetarian, or more precisely, a vegan diet, is an essential first step towards a compassionate relationship with the world. That relationship has to be grounded in integrity. We can no longer pretend to be compassionate while letting the farmer and the butcher do their dark deeds in secret. The consequences of such blatant subterfuges are clearly impacting us all on the planet.

It is not that hard to re-imagine our relationship with the world. One day, I was babysitting our granddaughter, Kimaya, and she asked me to take her to see the Disney movie, *Cinderella*[63]. I thought that I might be mildly amused by the familiar story, but I woke up when I heard the lead character articulate precisely what Climate Healers has been about. She said three simple sentences that I call the Cinderella Principles:

1. **Have Courage, Be Kind and All Will Be Well.** Courage is the original virtue, for without courage no other virtue can be exercised reliably. This is why Lord Krishna spoke to Arjuna in the Bhagavad Gita, for he among all the characters in the Mahabharata, exemplified courage. The evidence is clear that if we screw up our courage to be truly kind to all Life, then not only will the environmental crises abate, but also the planet will literally heal itself.

2. **Just Because It Is What Is Done, Doesn't Mean That It Is What Should Be Done.** If we don't question our ingrained habits, then how can we truly make radical changes in our relationship with the world? Just because we have been enslaving and abusing animals for the past 200,000 years, doesn't mean that we should continue to enslave and abuse them today.

3. **Imagine the World As It Should Be (and Work Towards It), Not the World As It Is.** Do you imagine a world thriving with life, where climate change and the other environmental catastrophes have been mitigated? Then that world will materialize only if we have the courage to act as if we want to make that world happen.

We have the power to change our relationship with the world just through our daily actions. The Miglets are already leading this change!

For the Caterpillar has no choice but to become a Butterfly.

# 6. Why Are We Here?

*"The farther backward you can look, the farther forward you are likely to see"* - Sir Winston Churchill.

In Nature, both the Caterpillar and the Butterfly serve useful purposes and have their respective niches in ecosystems. But in order to understand the purpose served by human beings in both our Caterpillar and Butterfly phases, we must examine the human story in a broader context.

Father Thomas Berry, a Catholic priest of the Passionist order, liked to call himself an Earth Scholar. He was a deep ecologist in the footsteps of Pierre Tielhard de Chardin and he considered the Universe to be the primary self-referential reality in the material world. Father Thomas Berry said that[1],

"The Earth is not part of the human story, the human story is part of the Earth story...The problem is we have tried to tell the human story without telling the Earth's story."

Therefore, in order to understand why we are here, our larger purpose, we need to tell the Earth's story. By that same logic, the Sun is not part of the Earth's story, but the Earth's story is part of the Sun's story. Therefore, if we want to understand the Earth's story, we need to tell the Sun's story as well.

## 6.1 The Earth's Story

Who or what is the Earth anyway? Is the Earth just a big lump of rock with useful resources on it, a "Blue Marble" as seen from outer space[2]? Or is the Earth a more complex material object governed by the laws of physics and chemistry, but still insentient? Or is the Earth also an intelligent, sentient being endowed with some independent agency executed through Life, though confined to a prescribed orbit around the sun, as in James Lovelock's "Gaia" model[3] and in the "Mother Earth" formulation embraced by most Eastern and indigenous wisdom traditions?

If our story is that the Earth is just a big blue marble, then of course, we would deny that the Earth could possibly be reacting to human activities with climate change, wildfires, droughts, floods, famines, ice melts and so forth. After all, we can scratch up a marble, throw it around and it never retaliates. Those who subscribe to such a story would be immune to scientific arguments about how physics, chemistry, geology and paleontology are all pointing to human activities causing climate change and environmental degradations on Earth. They would prefer to think that the sun, cosmic rays and other celestial forces directly cause any changes in the Earth's climate. As the Fox News TV personality, Bill O'Reilly, said[4],

> "Humans can't change the Earth's climate! Only God changes the Earth's climate!"

That ends any debate on climate change with those who subscribe to that view!

But if our story were that the Earth is a more complex material object governed by the laws of physics and chemistry, but nothing more, then we would deny that Life could possibly be involved in regulating the Earth's climate. True, trees absorb solar energy and transpire water through their leaves so that forests as a whole act like a green land surface with an evaporation channel for ground moisture. Climate models, or rather Global Circulation Models (GCM), can and do easily account for these physical characteristics, but that's where they stop. The statistician, George Box, said that,

> "All models are wrong. But some are useful."

As such, GCMs are indeed wrong, but they are undoubtedly useful. Unfortunately, climate scientists are particularly sensitive to any suggestion that their GCMs are wrong. They have been hounded on every little discrepancy in their scientific work and any such suggestion would be pounced upon by contrarians eager to promote fossil fuel interests[5]. Therefore, it was a triumph of academic integrity and the scientific method for the seminal paper on the Biotic Pump theory to be published in the Atmospheric Physics and Chemistry Journal of the European Geosciences Union[6]. It took two years of reviewing and revising, but it finally got published.

As the principal authors of the paper, Dr. Anastassia Makarieva and Prof. Viktor Gorshkov, describe it, the standard atmospheric GCMs used in climate modeling fail to account for half the rainfall over the Amazon forest. In the GCMs, the rainfall over the Amazon decreases linearly with distance from the ocean, whereas in reality, the rainfall over the Amazon is fairly constant and independent of this distance. One would think that is such a glaring discrepancy that would cause scientific heads to get scratched! But none of the GCMs based on physics and chemistry could account for that discrepancy. Apparently, we need biology to account for that!

In the standard GCMs, the total precipitation that falls on a region is determined by wind patterns, which are strictly due to temperature differentials. In the standard equation of eco-hydrology,

$$P = E+R,$$

where P is the precipitation, E is the evaporation and R is the runoff, it is assumed that when forests are cut down, the evaporation, E, declines and most of the precipitation, P, is available as runoff, R, which can then be dammed and used for human consumption. That is, the less forests there are the better it is for fresh water availability for humans!

But that is so contrary to common sense! If we ask any villager in India, she would tell us that when the forests are cut down, the rainfall decreases, water sources dry up and desertification begins. The total precipitation, P, tends towards zero and our grand plans to dam the runoff for human use come to naught. Therefore, something is amiss in the standard GCMs.

That something is filled in by the Biotic Pump theory.

In the Biotic Pump theory, forests also act to create their own rainfall. Along with the transpiration of water, forests emit microorganisms that become the nuclei for raindrop formation. As raindrops form above the forest, atmospheric pressure drops above the forest and a pressure gradient is created in the lower atmosphere, which causes moist air from above the ocean to be drawn towards the forest, and an atmospheric circulation ensues. Conversely, as forests are cut down, there would be more raindrop formation above the ocean than above land, which would cause the pressure gradient to reverse and result in moisture from land to be drawn over the ocean, leading to desertification. It is this atmospheric circulation due to Life that is completely missing in any climate model today. This atmospheric circulation due to Life is necessary to explain the rainfall over the Amazon. It also provides a plausible explanation for the steady expansion of the Sahara desert over the past 10,000 years.

But this means that the Earth's climate is not just a function of physics and chemistry, but Life regulates it as well!

Here's another situation where Life changes the environment to make it more hospitable for Life. In his engaging presentation at the 2009 AGU Annual Fall Meeting entitled, "*The Biggest Control Knob: CO2 in Earth's Climate History,*" Dr. Richard Alley, a Professor of Geosciences at Penn State University, described the $CO_2$ rock weathering thermostat to explain how the Earth's climate is regulated by the $CO_2$ level in the atmosphere[7]. The explanation is that of a purely chemical process involving volcanic activity and volcanic rocks that contain calcium and magnesium, which naturally absorb $CO_2$ to become limestone or dolomite. If the surface temperature of the Earth increases, the chemical reactions absorbing $CO_2$ increases and the atmospheric $CO_2$ is drawn down reducing the greenhouse effect, thereby lowering the surface temperature. If the surface temperature of the Earth decreases, the $CO_2$ draw down through rock weathering slows, building up the $CO_2$ levels in the atmosphere as volcanoes belch the gas out, thereby increasing the temperature. The time constant of this CO2 rock-weathering thermostat is on the order of 500,000 years.

Enter Dr. Ronald Dorn, a Professor of Biosciences of Arizona State University, who over the past quarter century, along with his students, has been studying how ants, termites and tree roots affect this rock weathering process[8]. At eight test sites in Arizona and Texas, he showed that ants could accelerate the absorption of atmospheric CO2 by 335 times compared to the chemical rock weathering process alone. Ants or perhaps, the microbes that coexist with ants are powerful biotic agents for the thermal regulation of the planet, just as forests are powerful biotic agents for atmospheric circulation.

Thus, Life is an active participant in regulating the Earth's climate and surface temperature and thereby creating the environmental conditions that promote the well being of Life. That should not be surprising given that each of us is endowed with biological mechanisms for thermal regulation and healing in our respective bodies. It is only fitting that the Earth who spawned us all possesses similar biological mechanisms to create an environment hospitable for Life, on a large, macro scale!

Therefore the time has come for us to accept the views of many Eastern and indigenous cultures, which consider the Earth to be a conscious being in her own right. Said a Smohalla representative in response to European entreaties in the 1880s[9],

"You ask me to plow the ground! Shall I take a knife and tear my mother's breast? Then when I die she will not take me to her bosom to rest.

You ask me to dig for stone! Shall I dig under her skin for her bones? Then when I die I cannot enter her body to be born again.

You ask me to cut grass and make hay and sell it, and be rich like white men! But how dare I cut off my mother's hair."

In his formulation, Mother Earth is indeed a sentient being in her own right, and we are one of her numerous children! Such a sentient planet orbiting an ordinary star, the Sun, which is like 90% of the stars in the Milky Way galaxy, has now spawned a pesky, terra-forming, superlative tool-building species that overspreads the globe and has the capacity to cause a major extinction event all by itself! It is unimaginable that the evolutionary process that spawned the perfection of the forest at the SAI Sanctuary also contains within it a suicidal flaw in the natural selection of superlative tool-building abilities. In one of his many inspiring passages, Father Thomas Berry wrote[10],

> "If the dynamics of the universe from the beginning shaped the course of the heavens, lighted the sun and formed the Earth, if this same dynamism brought forth the continents and seas and atmosphere, if it awakened life in the primordial cell and then brought into being the unnumbered variety of living beings, and finally brought us into being and guided us safely through the turbulent centuries, there is reason to believe that this same guiding process is precisely what has awakened in us our present understanding of ourselves and our relation to this stupendous process. Sensitized to such guidance from the very structure and functioning of the universe, we can have confidence in the future that awaits the human venture."

Indeed, not only can we have confidence in the future that awaits the human venture, but we can also have confidence in the past and present as well!

## 6.2 The Human Story

In order to comprehend why Life has spawned a tool-building species, we first have to embed the human story within the Earth's biography.

The Earth was formed by the accretion of the solar nebula about 4.5 billion years ago. While liquid water condensed on the surface of the Earth soon after, the first life forms on Earth appeared 3.5 to 3.8 billion years ago. But these were simple single-celled organisms. Complex ecosystems containing plants, animals and insects did not form until tens of millions of years after the Cambrian explosion occurred about 540 million years ago. Since that explosion, the biodiversity of species has been increasing almost monotonically and lately, almost exponentially, but interspersed with minor and major extinction events sprinkled throughout the fossil record[11]. Of these extinction events, five major extinction events were identified in a landmark paper in 1982[12]:

1) **The Ordovician-Silurian extinction event:** 450-440 Million years ago (Ma) at the end of the Ordovician period. It is speculated that two separate events occurred that killed off 60-70% of all species that existed at that time.

2) **The Late Devonian extinction event:** 375-360 Ma at the end of the Devonian period. It is estimated that this extinction event occurred over a prolonged period of about 20 Million years, with as much as 70% of all species dying out in that interval.

3) **The Permian Triassic extinction event:** 251Ma at the end of the Permian era. This is known as the Great Dying with as much as 90% of all species dying out. There is evidence that the atmosphere was full of hydrogen sulfide at that time, most likely caused when a climactic warming upset the oceanic balance between deep water sulfate reducing bacteria and photosynthesizing plankton. Hydrogen sulfide is not only poisonous to both marine life and life on land, it also weakened the ozone layer exposing life-forms to UV radiation as well.

4) **The Triassic-Jurassic extinction event:** 200Ma at the end of the Triassic era. It is estimated that 70-75% of all species went extinct and this event led to the rise of the dinosaurs.

5) **The Cretaceous-Paleogene extinction event:** 66Ma at the end of the Cretaceous era. This event led to the extinction of all non-avian dinosaurs and the rise of the mammals. The prevalent theory is that numerous celestial objects, mostly asteroids,

bombarded the Earth over a 300,000 year period and this resulted in the extinction event.

While these extinction events stand out, there are tens of other mass extinction events in the paleontological record and there are numerous hypotheses on potential causes for these extinction events. Identifying specific causes for particular mass extinction events is one of the most interesting detective tasks in the field of paleontology. The proposed cause should explain why the specific species groups died out while others survived the event and should be based on corroborating evidence that the conjectured events actually happened. In general, most mass extinction events can be traced to sudden climactic changes, either a sudden warming or a sudden cooling of the Earth's global surface temperature. The most commonly suggested causes of these sudden climactic spikes are as follows[13]:

1) **Sustained volcanic eruptions:** Sustained volcanic eruptions produce dust and sulphur dioxide ($SO_2$) emissions that block sunlight and therefore, impede photosynthesis and cause a cooling spike in the Earth's global surface temperature. The $SO_2$ rains as sulphuric acid after reacting with water vapor in the atmosphere, killing vegetation. Furthermore, sustained volcanic eruptions build up $CO_2$ in the atmosphere which means that once the eruptions stop, the dust settles and the $SO_2$ aerosols are removed from the atmosphere within a few days, the built-up $CO_2$ which lasts in the atmosphere for hundreds of years, causes a sudden increase in the Earth's global temperature resulting in the mass extinction event.

2) **Impact events:** The impact of a sequence of sufficiently large celestial objects, like the ones that famously killed off the dinosaurs 66 Ma, could have similar consequences as sustained volcanic eruptions. Such impacts could result in global wildfires, acid rain and global heating due to large $CO_2$ build-up in the atmosphere. It is believed that a large asteroid impacting the ocean could have worse consequences for life than one impacting land, since a local heating of seawater over 50oC due to such an impact can result in a huge spike in $CO_2$ emissions from the ocean.

3) **Clathrate Gun:** A warming of the Earth's global surface temperature can thaw large amounts of frozen methane (methane that is stored in a lattice of ice) that is normally found in the Earth's colder surfaces. Since methane is a much more powerful greenhouse gas than $CO_2$, this can become an amplifying feedback since the methane emissions would increase the temperatures further.

4) **Gamma ray bursts:** A nearby supernova or gamma ray burst in the galaxy, within 6000 light years of the solar system, could cause the ozone layer to be weakened exposing life to fatal levels of UV radiation. But gamma ray bursts are rare, occurring only a few times in the Milky Way galaxy every million years, though it is suggested that a gamma ray burst was one of the events responsible for the End Ordovician extinction that occurred 450-440 Ma, just as complex life was beginning to evolve on Earth.

Other than the fourth type of event, which is extremely rare, the effects of the other events are similar to what is occurring on Earth today due to human industrial activities. Humans have deliberately pumped greenhouse gases and aerosols into the atmosphere at a pace comparable to what would have occurred if a sustained volcanic eruption had happened. Humanity is being forced to perfect methodologies that would draw down these atmospheric gases on Earth reliably, on an ongoing basis. Also, in the process of becoming a global industrial civilization, humans have developed powerful telescopes, space travel technologies and nuclear weapons technologies that make it possible to combat any barrage of large asteroids, comets and meteors from ever impacting the Earth again.

Therefore, I contend that Mother Earth's purpose of spawning of a superlative tool-building species was to combat the primary causes of major extinction events that had happened on Earth! It's as if Mother Earth inoculated herself with a vaccine, our species. True, the vaccine has caused a mild fever and a mild extinction event that is gathering momentum, just as a medical vaccine in a human being is prone to cause mild symptoms of the disease that she's being inoculated against. But both of these symptoms can be ameliorated if humans begin taking the right decisions now, once we seriously assume our responsibilities in our ecological niche within the planet's ecosystems - as the "sentinel caregivers," the *"Khalifahs"* for all Life. As we assume this role in our Butterfly phase, we would be like the prairie dog sentinel in the prairie dog colony that is the Earth's biosphere.

Imagine the following news item, dated Sep. 8, 2015:

> "Discovered on August 24, 2015, by the Lincoln Near Earth Asteroid Research Project, an MIT Lincoln Laboratory program funded jointly by NASA and the US Air Force, in New Mexico, the asteroid code named 2014 QQ47 has been classified as a 10 on a Torino scale of impact hazards, the highest threat level possible. The orbit of this asteroid has been calculated using 51 observations over a 2 week period and scientists are 95% certain that it is headed for a collision with the Earth on March 21, 2026. If this asteroid strikes the Earth intact, the impact will have the effect of over 80 million Hiroshima style atomic bombs. As Billy Bob Thornton says in Armageddon, "It's what we call a Global Killer....the end of mankind. Half the world will be incinerated by the heat blast.....the rest will freeze to death in a nuclear winter. Basically, the worst part of the Bible!"

The White House issued a press release stating that President Obama is in emergency meetings with NASA scientists, military officials and world leaders to devise a strategic response to the asteroid threat. He urged the American people and the people of the world to stay calm and rest assured that the world's military powers have the technological capabilities to defend the Earth from this cataclysmic threat.

Scientists from the NASA Jet Propulsion Laboratory in Pasadena, California, stated that the sooner we begin a crash space program to transport and detonate a large thermonuclear device on the asteroid, the better. But even if such a detonation is successful, there is a strong likelihood that tens of thousands of large asteroid debris would rain on Earth on March 21, 2026 possibly requiring an unprecedented effort involving thousands of military personnel tracking down and shooting the larger debris with missiles before they impact the Earth."

Just imagine the unprecedented cooperation among people and nations that would occur when faced with such an external, existential threat. Gone would be all our differences over race, color, creed, national boundaries, sexual orientation or whatever utter triviality that divides us these days. Gone would be our self-doubt over our preoccupation with technology, especially military technology, over the past 200 years. We would even be so universally thankful that Americans pioneered the development of nuclear weapons and space travel technologies over the past century. There is no doubt that we will come together as one humanity to respond, to protect the Earth, not just for our sake, but for the sake of all our fellow Earthlings. We would then have discovered our true identity as a species, as a sentinel caregiver for all Life on Earth.

This is not such a fanciful threat, either. The first paragraph in the above imaginary news item appeared almost verbatim in a news item in 2003, except that I changed the dates by 12 years and I changed the threat level on the Torino scale from 1 to 10[14]. Subsequent observations and refinement of orbit calculations of the asteroid 2003 QQ47 downgraded the threat level on the Torino scale to 0, a false alarm[15].

In the same year, 2003, a scientific paper published in the Proceedings of the American Institute of Physics settled the question[16]:

"Preventing collisions with the Earth by hypervelocity asteroids, meteoroids, and comets is the most important immediate space challenge facing human civilization. This is the Impact Imperative. We now believe that while there are about 2000 earth orbit crossing rocks greater than 1 kilometer in diameter, there may be as many as 200,000 or more objects in the 100 m size range. Can anything be done about this fundamental existence question facing our civilization? The answer is a resounding yes! By using an intelligent combination of Earth and space based sensors coupled with an infra-structure of high-energy laser stations and other secondary mitigation options, we can deflect inbound asteroids, meteoroids, and comets and prevent them from striking the Earth. This can be accomplished by irradiating the surface of an inbound rock with sufficiently intense pulses so that ablation occurs. This ablation acts as a small rocket incrementally changing the shape of the rock's orbit around the Sun. One-kilometer size rocks can be moved sufficiently in about a month while smaller rocks may be moved in a shorter time span."

But why did the Earth spawn such a sentinel caregiver species now? Clearly, there have been numerous extinction events in the past 500 Million years of the Earth's history, during which complex life evolved and indeed, biodiversity thrived as a result of these extinction events[17]. Therefore, what makes this era so critical to demand the presence of such a species? To understand why the Earth deems this so critical, we need to consider the evolution of the sun and the history of atmospheric CO2 on Earth.

## 6.3 The Sun's Story

Our solar system orbits the black hole at the center of the Milky Way galaxy with a period of 225Ma-250Ma, the Galactic Year or the Cosmic Year[18]. This means that approximately one Galactic year ago, the Permian-Triassic Extinction event or the "Great Dying" occurred on Earth, with about 90% of all species going extinct. We are entering the first galactic anniversary of the P-T extinction event, the Great Dying, as far as the solar system is concerned.

Our sun is also a G-type main sequence star (G2V) and assuming that it didn't exhibit strange anomalies during its evolution, it would have had a fairly predictable increase in intensity as its hydrogen molecules get converted to helium in the nuclear fusion reactions in its core. Since birth, the luminosity of the sun would have increased by about 40% and over the past 500 million years during which complex life evolved, the luminosity of the sun would have increased by about 5%[19]. While a luminosity increase of just 5% may seem like a minor change, it actually has major implications for the sensitivity of the Earth to greenhouse gas perturbations. Greenhouse gases are like the blanket that the Earth wears to ensure optimal conditions for Life and as the sun gets hotter, the Earth would need to shed those blankets.

When the solar system was born 4.6 billion years ago, planet Earth was perfectly situated near the center of the Habitable Zone (HZ) around the Sun. The HZ is the zone where liquid water would be available on the surface of the planet, which is essential for life to flourish on the planet. The boundaries of the HZ are typically calculated using 1-dimensional, cloud-free climate models as pioneered by Kasting et al[20]. The inner edge of the HZ is calculated for the loss of liquid water on the surface and the outer edge of the HZ is calculated using the maximum greenhouse effect. Using this model, it can be shown that the HZ around our sun, at present, is 0.95AU to 1.67AU, where AU is the Astronomical Unit, the average distance of the Earth from the Sun. In this model, the Earth is presently 0.05AU away from the inner edge of the Habitable Zone around the sun.

Recently, an updated 1-Dimensional, radiative-convective, cloud free climate model, with updated $H_2O$ and $CO_2$ absorption coefficients was used by Dr. Ravi Kopparapu and others at Penn State University to show that the HZ limits of our solar system is from 0.99AU to 1.70AU[21]. This would mean that the Earth is extremely close to the inner edge of the Habitable Zone and therefore, very sensitive to perturbations in the greenhouse gas composition in the atmosphere. According to best estimates, the $CO_2$ concentration in the Earth's atmosphere 500 million years ago when complex life first formed on Earth was between 4000ppm to 20,000ppm. In that atmospheric environment, a 2000ppm increase in $CO_2$ concentration would, at most, be a 50% increase, a similar percentage increase to the one that we've caused in the Earth's atmosphere since pre-industrial times. However, the same 2000ppm increase in $CO_2$ concentration from pre-industrial levels would be like a 700% increase and most likely fatal to Life on Earth today[22]. This is because the $CO_2$ concentration in the pre-industrial atmosphere was a relatively minuscule 280ppm.

In such an extremely sensitive atmospheric environment, at the very inner edge of the Habitable Zone around the Sun, the Earth spawned a superlative, tool-building species that has the capacity to quickly reverse any major perturbations in the atmospheric composition through technological interventions. Knowing that the chemical feedback mechanisms were too slow to prevent catastrophic consequences to all Life at this inner edge of the Habitable Zone around the sun, the Earth fashioned some fast feedback mechanisms through Life - to defend Life. In order to ensure that our species fulfills the caregiver role in the ecosystem, the Earth birthed our species with the core Dharma of compassion for all creation. Further, to ensure that the species develops the necessary technology first to fulfill its sentinel role, the Earth spawned the species in an environment of abject terror so that the species overcomes its core Dharma and spends 200,000 years fashioning increasingly sophisticated weapons and technologies to overcome that terror. In the process, the species creates an atmospheric perturbation similar to, though not as severe as that which would be encountered with sustained volcanic eruptions or through an asteroid impact in the ocean. In that sense, what we're about to embark upon in the next few decades is a trial run for the real long-term tasks that we've been spawned to do.

The Earth had to force us to go through all that violence and suffering of the past 200,000 years for there is no way that a human society full of compassionate, enlightened "Buddhas" would have ever been motivated to develop sophisticated weapons technologies and constantly seek to improve those technologies. It is only fratricidal, intra-species warfare that could have compelled us to develop these technologies. It is only natural that when we are in a system that has normalized war, the act of killing other human beings, we would consider it normal to destroy forests and the ocean and kill other species. But it is mostly the same sophisticated technologies that are now being deployed to monitor the Earth's biogeophysical and biogeochemical processes. In the future, it is these same technologies that will be deployed to defend all Life on Earth from both internal and external threats.

Thus Mother Earth, or if you will, God, Yahweh, Allah, Brahman, the Creator, the Great Spirit, Higher Consciousness, or just plain Life, that beautiful process which seemingly creates order out of chaos, with infinite compassion for all Life, spawned a species that can control the Earth's $CO_2$ thermostat more finely, deter the collisions of asteroids and comets, and protect Life against any existential threats, external and internal, even as the Sun inexorably gets brighter over time and as the solar system enters the first galactic anniversary of the Great Dying. Who knows, in another 500 million years, when the Sun is 5% brighter than today, the evolutionary descendants of our species will deploy technologies to reflect all that excess energy from the Sun back into space and allow life to continue on Earth, well beyond the inner edge of the Habitable Zone.

The possibilities are endless, when our technological prowess is marshaled in the service of Life. As Father Pierre Tielhard de Chardin predicted nearly a century ago[23],

"The day will come when, after harnessing space, the winds, the tides, gravitation, we shall harness for God the energies of love. And, on that day, for the second time in the history of the world, man will have discovered Fire."

But in our story, we have always been harnessing "for God the energies of love," for we have always been building our tools and technologies for a larger purpose. We unknowingly stabilized the Earth's climate during the first 10,000 years of the Holocene era, mainly through deforestation and desertification. But over the past 200 years, we developed all the technologies we need to consciously maintain that climactic stability for the foreseeable future, without having to depend on further deforestation and desertification. In the process, we have caused the Earth to suffer a mild fever, but we know that this fever will subside once we help regenerate forests on half the Earth's land surface and we transition from fossil fuels to clean energy sources.

Besides, an Earth with a stable climate maintained by a compassionate human presence, is an Earth that no longer has to undergo the periodic glaciations and deglaciations that occurred over the past 3 million years due to the Milankovitch cycles. Every such climactic gyration caused untold suffering to species that had to migrate and compete for habitat. Therefore, in the final analysis, the temporary suffering that humans and other species had to endure as we developed our tools and technologies, will turn out to be minor compared to the suffering that is avoided with our compassionate, technological presence.

Therefore, we are no different from the elephants of SAI sanctuary that so effortlessly and routinely contribute to the good of the planet's biosphere. We are no different from the forests that regulate rainfall, or the ants that sequester $CO_2$. We are the thermostat species, the sentinel caregivers (*Khalifahs*) of the Earth!

In that affirmation of a plausible evolutionary purpose grounded in science, just as envisioned in the world's religious scriptures, we will indeed be rediscovering Fire for the second time!

# 7. Everything Will Change

"*Imagination is more important than knowledge. For knowledge is limited to all we now know and understand while imagination embraces the entire world, and all there ever will be to know and understand*" - Albert Einstein.

Wayne Dyer, the noted self-help author and motivational speaker once said[1],

"When you change the way you look at things, the things you look at change."

That is precisely what our story does by reversing the mainstream perspective. Instead of humans controlling Nature, in our story, Nature had been controlling humans all along, putting us to work developing the tools and technologies that were necessary to create a fast feedback mechanism for prolonging Life at the inner edge of the Habitable Zone. Therefore, Jean-Jacques Rousseau's wise words from the Social Contract applies to us at a species level[2]:

"Man is born free, and everywhere he's in chains. One believes himself the master of others, who nevertheless is more a slave than they."

In our story, humans were allowed to commandeer Nature's treasures, including fossil fuels and other species, but on a transient basis in our tool building, Caterpillar phase. Though we enslaved animals en masse, we were really more slaves than they, especially in our technological society. We were wage slaves, debt slaves or just plain slaves to our own ambitions and desires. The more we tried to control Nature to produce monocultures to our liking, the more super weeds, super bugs and zoonotic diseases that Nature created to keep us frozen in a high state of anxiety. Now, we're receiving signals from Nature that our frenzied tool-building phase is over and that it is time to free ourselves from our invisible chains, repair the collateral damage that has occurred and heal the planet and ourselves.

HEAL is literally an acronym for Human, Earth and Animal Liberation. For a sustainable human presence on Earth would be truly liberated and nonviolent, in all respects.

This formulation of the human story fits the known facts just as well as the typical "gloom and doom" environmental stories, while putting the trials and tribulations of our fellow beings and our ancestors in a different light. Our fellow beings have suffered tremendously at our hands over the past 200,000 years. It is now time for us to ease their suffering and nurse them back to health. Our ancestors were like sculptors working on various aspects of a multi-generational, long-term, technological project. It is only now that we can begin to see the fruits of their endeavors come together. While many realized souls among our ancestors recognized the misery and suffering in our enslaved human condition and even showed us how to free ourselves, an inner compulsion drove us to pay lip service to their teachings and complete this multi-generational project. For the good of all Life!

Today, many of us have been feeling divided, dejected and depressed by our socioeconomic and environmental crises. But in our story, there are no billionaires, foreigners or "freeloaders" to blame for the difficulties that we face today. It is only when we stop assigning blame for our perceived difficulties that we can come together to do the needful - a radical transformation of our socioeconomic system, from consumption to compassion as an organizing value, and from competition to collaboration as an organizing principle.

Just over a hundred years ago, Mahatma Gandhi faced the same situation in his beloved India - a people divided, dejected and demoralized by their British colonial rulers. Gandhi wrote his famous monograph, *Hind Swaraj* or *Indian Home Rule*[3] in 1909, to perk up his countrymen and instill in them a sense of pride for their culture and traditions. The British colonial rulers of India banned the book, which naturally made it an instant best seller! In the book, Gandhi addressed the prevailing narrative of the colonizers that Indians were intellectually inferior since they weren't leading the scientific and industrial revolution in the world. He pointed out that true happiness is only rudimentarily dependent on the material comforts that the scientific and industrial revolution was enhancing at the expense of our spiritual disconnection and therefore, Indians should be teaching others, rather than learning from others:

"Civilization is that mode of conduct which points out to man the path of duty. Performance of duty and observance of morality are convertible terms. To observe morality is to attain mastery over our mind and our passions. So doing, we know ourselves. The Gujarati equivalent for civilization means "good conduct".

If this definition be correct, then India, as so many writers have shown, has nothing to learn from anybody else, and this is as it should be. We notice that the mind is a restless bird; the more it gets the more it wants, and still remains unsatisfied. The more we indulge our passions the more unbridled they become. Our ancestors, therefore, set a limit to our indulgences. They saw that happiness was largely a mental condition. A man is not necessarily happy because he is rich, or unhappy because he is poor. The rich are often seen to be unhappy, the poor to be happy. Millions will always remain poor. Observing all this, our ancestors dissuaded us from luxuries and pleasures. We have managed with the same kind of plough as existed thousands of years ago. We have retained the same kind of cottages that we had in former times and our indigenous education remains the same as before. We have had no system of life corroding competition. Each followed his own occupation or trade and charged a regulation wage. It was not that we did not know how to invent machinery, but our forefathers knew that, if we set our hearts after such things, we would become slaves and lose our moral fiber. They therefore, after due deliberation decided that we should only do what we could with our hands and feet. They saw that our real happiness and health consisted in a proper use of our hands and feet. They further reasoned that large cities were a snare and a useless encumbrance and that people would not be happy in them, that there would be gangs of thieves and robbers, prostitution and vice flourishing in them and that poor men would be robbed by rich men. They were, therefore, satisfied with small villages. They saw that kings and their swords were inferior to the sword of ethics, and they, therefore, held the sovereigns of the earth to be inferior to the Rishis and the Fakirs. A nation with a constitution like this is fitter to teach others than to learn from others."

With that positive formulation of the Indian's story, he was able to transform Indian attitudes and thus lead the movement to free India from British colonial rule.

## 7.1 The Signs of the Metamorphosis

Throughout his life, Gandhi urged Indians and the whole world to adopt a simple lifestyle for the true pursuit of happiness, for such pursuit is entirely inward looking once basic necessities are met. Looking back, we now see that Gandhi's harsh assessment of Western civilization in *Hind Swaraj* was premature. But he couldn't have known at that time what all these Western obsessions over money, machinery and material progress were really about. Indeed, imagine if the world had heeded Gandhi's words then and halted the Military Industrial Complex that was fueling Western civilization throughout the 20th century. Then there would be no Internet, no nuclear weapons, no space technologies, no computers and, consequently, no defenses against asteroids, comets or prolonged volcanic eruptions. All these essential technological developments and scientific discoveries occurred within a century at such a breathtaking pace due to the singular Western belief, however false, that continuous economic growth, together with a competitive, militaristic culture steeped in consumerism, is necessary for the pursuit of happiness.

Such is the paradox of perfection. The perfection of the whole arises despite the blatant "flaws" in the individual elements. As the spiritualist and philosopher, Ram Dass, said[4],

"It's all perfect, and it all stinks. The conscious being lives simultaneously with both of these."

A strange paradox, indeed!

Currently, the industrial world is embedded in a hierarchical socioeconomic system that was optimally designed for the tool-building Caterpillar phase of our species. This system is really just a huge military-industrial undertaking, with animals, workers, smelters and miners, along with clerical, administrative, marketing, executive and financial personnel. Think of this system as a vast pyramid, where animals and Nature occupy the lowermost layer, with humans layered above in the order of increasing rank privileges. These privileges are assigned on the basis of merit or ancestry or cultural or physical attributes, such as race, color, creed, caste, gender or sexual orientation. There are educational privileges, male privileges, skin color privileges, caste privileges, heterosexual privileges and class privileges, to name a few. The pyramid is fueled by energy and capital in the form of money, poured in at the top, issued through debt. The money is then fanned out to the lower layers with the requirement that the principal plus interest must be returned back to the top. Since there is no currency issued to cover that interest, it creates a game of musical chairs and therefore an environment of scarcity and competition. This provides a strong impetus for the pyramid to grow with new money (i.e., debt) being issued continuously in order for this economic system to work. As it grows, the pyramid increases its footprint on Nature, destroying wildlife habitats and forcibly displacing or assimilating indigenous communities at its base. This is a system rooted in violence, but optimally designed for its militaristic purpose!

This system spreads misery and suffering to every participant. The animals at the very bottom of this pyramid are brutally enslaved, oppressed and slaughtered. The humans right above the animals are the miners, the smelters, the farm workers, etc., while those in the middle layers perform the administrative and clerical functions and those in the top layers take on the executive, marketing, financial, research and development responsibilities. While the misery in the lower layers occurs in the form of poverty, hunger, physical oppression and diseases of scarcity, the misery in the upper layers occurs in the form of mental anguish and diseases of excess such as cancer, heart disease, stroke and diabetes. Indeed, the suffering can be even more acute in the uppermost layers than in the lower layers due to the extreme isolation that people experience.

Denials, deceptions and coercion are like the nuts, bolts and glue that hold this pyramid together. The many myths that sustain this system include cultural myths that extol the superiority of the industrial civilization over indigenous ways of living, sexist myths that extol the superiority of men over women, racist myths that extol the superiority of the white race over people of color, homophobic myths that extol the superiority of heterosexuals over LGBT people and above all, speciesist myths that extol the superiority of human beings over all other creatures.

The hierarchical organization of the socioeconomic system, by itself, fosters myriad oppressions. In the documentary, *Zeitgeist: Moving Forward*, Peter Joseph observes[5],

"Make no mistake: the greatest destroyer of ecology, the greatest source of waste, depletion and pollution, the greatest purveyor of violence, war, crime, poverty, animal abuse and inhumanity, the greatest generator of social and personal neurosis, mental disorder, depression, anxiety, not to mention, the greatest source of social paralysis, stopping us from moving into new methodologies for personal health, global sustainability and progress on this planet, is not some corrupt government or legislation, not some rogue corporation or banking cartel, not some flaw of human nature and not some secret hidden cabal that controls the world. It is, in fact, the socioeconomic system itself at its very foundation."

But this system is clearly at its breaking point. This is why leaders from across the political spectrum in every nation are going to extreme lengths to maintain a continuous economic growth paradigm even in the face of global environmental catastrophes, because without such growth, its debt-based monetary system could collapse, just as any financial Ponzi scheme would collapse when there are no fresh recruits to contribute to it. No leader wants such a collapse to happen on his or her watch. As George Monbiot writes[6],

"To try to stabilize this system, governments behave like soldiers billeted in an ancient manor, who burn the furniture, the panelling, the paintings and the stairs to keep themselves warm for a night. They are breaking up the post-war settlement, our public health services and social safety nets, above all the living world, to produce ephemeral spurts of growth. Magnificent habitats, the benign and fragile climate in which we have prospered, species that have lived on earth for millions of years, all are being stacked onto the fire, their protection characterized as an impediment to growth."

It is not just governments. Upon the altar of growth, leaders of all stripes are beginning to blatantly deny various aspects of reality. The leaders on the political left selectively ignore any environmental signals that question the viability of a continuous economic growth paradigm. In his NY Times blog, The Conscience of a Liberal, the Nobel Laureate economist, Paul Krugman, claimed[7],

"Saving the planet would be cheap; it might even be free... [It] would have hardly any negative effect on economic growth and might actually lead to faster growth."

But, faster growth invariably entails more human consumption! As the economist, Charles Eisenstein, explained[8],

"What does economic growth actually mean? It means more consumption – and consumption of a specific kind: more consumption of goods and services that are exchanged for money."

Therefore, anything that reduces global consumption becomes anathema to leaders on the political left. This is why most mainstream environmental organizations advance the view that some form of grass-fed beef eating, solar-paneled, electric car driving continuous economic growth consumer culture can be spawned from the ashes of our current fossil-fueled consumer culture. Even the climate scientists of the UN IPCC go out of their way to deny that addressing climate change would have a major impact on our way of life or on economic growth. As Rajendra Pachauri, the former chair of the UN IPCC, said[9],

"We have the means to limit climate change. The solutions are many and allow for continued economic and human development. All we need is the will to change."

The latest UN IPCC report goes on to state[10],

"Tackling climate change need only trim economic growth rates by a tiny fraction, and may actually improve growth by providing other benefits, such as cutting health-damaging air pollution."

While making such statements, the UN IPCC is selectively ignoring the reality of species extinctions, habitat loss, nitrogen cycle disruptions, ocean depletion and other environmental alarm signals that are blaring for a contraction in our human footprint on the planet. In contrast, leading conservatives on the right appear to have understood that seriously tackling climate change would be incompatible with maintaining an ever-growing economy and therefore choose to clutch at any stray data that reassures them that climate change isn't happening or that it isn't human induced and therefore, might naturally reverse itself. Writes Jonathan Kay of the National Post[11],

"In the case of global warming, this dissonance is especially traumatic for many conservatives, because they have based their whole worldview on the idea that unfettered capitalism — and the asphalt-paved, gas-guzzling consumer culture it has spawned — is synonymous with both personal fulfillment and human advancement. The global-warming hypothesis challenges that fundamental dogma, perhaps fatally."

Therefore, leaders on all sides appear to be blind to the clear indications that everything that we consider normal today needs to change, that our material growth phase is over. But the views of these leaders percolate through the media to the masses. With respect to climate change, Erik Lindberg described the situation as follows[12]:

"We have a situation, then, where one half of the population says it is not happening, and the other half says it is happening but fighting it doesn't have to change our way of life. Like a dysfunctional and enabling married couple, the bickering and finger-pointing, and anger ensures that nothing has to change and that no one has to actually look deeply at themselves, even as the wheels are falling off the family-life they have co-created."

That is surely due to the cultural conditioning of the system that we're embedded in. After all, it is easy to articulate the behavioral changes that are needed to create a steady state version of the industrial civilization. The Cinderella Principles are not that hard to comprehend. As the 12-year old Severn Suzuki told the delegates at the UN Rio Summit in 1992[13]:

"At school, even in kindergarten, you teach us how to behave in the world. You teach us to not fight with others. To work things out. To respect others. To clean up our mess. Not to hurt other creatures. To share, not be greedy. Then why do you go out and do the things you tell us not to do? Do not forget why you are attending these conferences - who you are doing this for. We are your own children. You are deciding what kind of world we are growing up in.

Parents should be able to comfort their children by saying

"Everything's going to be all right. It's not the end of the world. And we're doing the best we can." But I don't think you can say that to us anymore. Are we even on your list of priorities? My dad always says "You are what you do, not what you say." Well, what you do makes me cry at night. You grown ups say you love us, but I challenge you, please make your actions reflect your words."

But her words fell on deaf ears. Later, after watching 20 years of global dithering on climate change, and as a 32-year old mother, Severn Cullis-Suzuki said at the UN Rio+20 summit[14],

"Twenty years on since Rio, we need nothing short of a massive paradigm shift to a strategic way of living that will carry our human race forward to a future with dignity."

The climate scientist, Dr. Kevin Anderson, made a similar assessment in 2013[15]:

"Perhaps at the time of the 1992 Earth Summit, or even at the turn of the millennium, 2°C levels of mitigation could have been achieved through significant evolutionary changes within the political and economic hegemony. But climate change is a cumulative issue! Now, in 2013, we in high-emitting (post)-industrial nations face a very different prospect. Our ongoing and collective carbon profligacy has squandered any opportunity for the 'evolutionary change' afforded by our earlier (and larger) 2°C carbon budget. Today, after two decades of bluff and lies, the remaining 2°C budget demands revolutionary change to the political and economic hegemony."

But such a revolutionary change to the political and economic hegemony would need revolutionary changes in our personal behaviors as well, along the lines that Severn Cullis-Suzuki articulated in 1992. After all, almost everything we do in our global industrial society, what we wear, what we eat, what we do and how we live harms the environment. Our per capita energy use is already so enormous that every human being has the equivalent of 22 energy slaves at his or her beck and call. This does not even include the energy embedded in the food we eat! In richer societies such as the US, it is on the order of 150 energy slaves per person. On average, Americans use more energy in a month today than most of their great grandparents used in their entire lifetime, just a century ago. But such profligate energy use levels were surely just a one-time splurge during the most productive, tool-building, technological growth phase of our global industrial civilization! In fact, more than 90% of our energy use is currently for wasteful, planet-destroying activities[16]. Just think that the non-profit Wikimedia foundation serves 500 million unique monthly visitors to the Wikipedia website with a staff of just 200, whereas the for-profit corporation, Google, employs 50,000 people for the 1 billion unique monthly visitors that Google serves. Those 50,000 Google employees are mostly engaged in the business of trying to persuade all of us to buy things that we don't really need[17].

There are other signals indicating that our material growth phase is over as well. Apple Computer, the largest corporation in the world in terms of market capitalization, was as usual, being very secretive about what its next major "iProduct" was going to be.

Will it be the "iWatch"?

Will it be the "iGlass"?

But did it really matter?

These gadgets are now more about titillation than necessity, even as we're being exposed, almost daily, to the sheer misery involved in producing these electronic products at such a frantic pace and gargantuan scale. There is misery in the mining of the minerals used in the products, misery in the assembly line at the factories and misery due to the chemical pollution during production, disposal and even recycling of these products. Meanwhile, the iPhone 6 does almost exactly the same things as the iPhone 5, but with a flimsier frame that easily bends. On the defense front, the US military is now seeking one trillion dollars to develop a new fighter jet, the F35, a blatantly unnecessary appendage to the already vast American military arsenal. President Obama has authorized spending another trillion dollars for a complete renewal of American nuclear weapons.

These are all indications that we have saturated in our military and technological needs as a species. That it is time to transition to a steady-state civilization, assume our identity as the sentinel caregivers for all Life on Earth, and awaken into our Butterfly phase.

Signs of that metamorphosis are all around us.

## 7.2 The How of Change

Dr. Vandana Shiva, the Indian humanitarian and environmentalist, said recently[18],

> "There are two great trends sweeping the world: one, a trend of diversity, democracy, freedom, joy, culture, people celebrating their lives. And the other, monocultures, deadness, everyone depressed, everyone on Prozac. We don't want that world of death."

It's fitting that at this time of great transformation, we see extreme polarization so that people can clearly discern the fork in the road. Which way do we want to go: evolve towards Utopia and careen towards Oblivion? In his memorable opus, *Why the West Rules For Now*[19], Ian Morris contended that this is precisely the choice we are facing today. There was a similar moment in the 1960s when the great American architect, systems theorist and inventor, Buckminster Fuller, authored the book, *Utopia or Oblivion*[20], in 1969, but Norman Borlaug and his Green Revolution postponed our day of reckoning for a few decades. Borlaug won a Nobel Peace prize for his efforts, but the trouble with postponing the day of reckoning is that the choices become so much starker, the next time around.

To reach a steady-state Butterfly civilization, we must nurture and accelerate the former trend of diversity, democracy, freedom, joy, culture, people celebrating their lives, while repurposing the technologies that we've developed to help life thrive on Earth. Think of the millions of Non-Governmental Organizations that have sprung up to address social ills of every kind. Think of the idealistic young men and women who are willing to take steep pay cuts in order to work on social causes. A Brookings Institution Report indicated that 64% of the Miglets would prefer to work in a meaningful job paying $40K per year as opposed to a meaningless job paying $100K per year[21]. This is the great trend that is leading towards the awakening of our Butterfly phase. As this trend is fostered, I imagine that we will be focusing our cameras on the forests of the world to monitor their regeneration, instead of focusing them on ordinary people for ubiquitous surveillance. I imagine that we will be building solar greenhouses yielding organic produce year-round in our urban gardens, instead of building cheap computer-controlled drones that drop more useless merchandise on our suburban front lawns. I imagine that we will be using our technological skills to build fewer, high quality, essential things that last instead of more cheap, low quality, disposable things that break.

The latter trend that Dr. Vandana Shiva mentions, of monocultures, deadness, everyone depressed, everyone on Prozac, is just the terrible hangover from our frenzied, tool-building Caterpillar phase. These are the symptoms indicating that our current hierarchical socioeconomic system is so corrupt that it is time to dissolve it. But this is unlikely to occur from the top down without a significant grassroots movement from the bottom up.

At the moment, mental health problems are rampant in the upper echelons of our hierarchical system with one in four people in the US suffering from depression and other mental diseases[22]. In one of the richest societies in the world, roughly half the American public consumes anti-depressants, anti-anxiety medications or illegal drugs on a daily basis, which is an indication of their immense isolation, misery and suffering[23]. Illicit drug use is pervasive up and down the economic scale in American society, with Wall Street executives ingesting more drugs on average than the poor people who disproportionately crowd federal prisons for drug abuse. Yet, they desperately cling to the power that they wield and the privilege that they think they have earned. Corruption is rampant at all levels of our world political structures with Americans, especially young Americans, beginning to tune out of the political process. Recent voting turnouts in midterm elections are in an abysmal 20%-40% range in most congressional districts in the US. The NY Times op ed columnist, Nicholas Kristof, wrote[24],

"Let's face it: The American political system is broken. The midterm elections were a stinging repudiation of President Obama, but Republicans should also feel chastened: A poll last year found Congress less popular than cockroaches."

Not only is the US Congress less popular than cockroaches, the same poll found Congress, which has an 8% approval rating, to be less popular than dog poop, hemorrhoids, toenail fungus, witches and mothers-in-law[25]. This is the same Congress that continues to foster the latter trend of monocultures and deadness to keep the current hierarchical system going. Two sociologists, Martin Gilens and Benjamin Page, have shown that while the US Congress is highly responsive to the interests of the economic elite, it is utterly indifferent to the interests of the general public[26]. The former NASA scientist, James D'Angelo, has traced this indifference to the so-called "Sunshine Act" of 1970, which essentially eliminated the secret ballot for Congressional votes[27]. If votes can't be cast in secret in any system, then intimidation and corruption becomes endemic. But this is not just an American phenomenon, as the governments of the world have been continuing to foster the latter trend of deadness as well, even as they meet annually in the UN Climate Change conferences. In a brilliant article entitled, "*Are we on the Verge of Total Self Destruction?*". Prof. Noam Chomsky of MIT wrote[28],

"At one extreme, you have indigenous tribal societies trying to stem the race to disaster. At the other extreme, the richest, most powerful societies in world history, like the United States and Canada, are racing full-speed ahead to destroy the environment as quickly as possible."

The indigenous tribal societies all have one thing in common: they consider the Earth to be sacred. Many of them had already fashioned steady-state civilizations in harmony with the Earth in their local environments before they were disrupted and dislodged by the ever-expanding base of the global industrial civilization. They were the "early adopters" in the trend of diversity, democracy, freedom, joy and culture that Dr. Shiva spoke about. The richest, most powerful societies in the world all have one thing in common: they are effectively ruled by transnational financial institutions and corporations who consider the Earth to be composed of resources to be processed for economic profit and material progress. They foster the latter trend of monocultures, deadness, everyone depressed, everyone on Prozac, that world of violence and death. Now the rich societies are using their immense military and economic power to crush the remaining indigenous tribal societies in the name of progress. Nurturing elements within the rich societies get caught up in this conflict. For they face a dilemma either submit to the dominant exploitative view of the Earth or face discrimination and oppression in the name of the greater good. In the rich societies, even grandmothers are branded as "eco-terrorists" and thrown in jail for daring to oppose this exploitative paradigm[29].

While the pervasive discrimination and oppression in the hierarchical system can be classified as economic colonialism, sexism, racism, etc., the underlying motivation is mostly just business. Economic colonialism reduces commodity prices. Sexism sells. Racism creates cheap prison labor and wage slaves. It is the debt paradigm of our world financial system and the quest for ever-growing profits that drives most injustices in the current system[30]. Any hierarchical system needs to select who gets to be on top and who languishes at the bottom, which fundamentally fosters such injustice and inequality. As Ashley Maier and Stacia Mesleh described it[31],

"A pervasive mindset, conscious or unconscious, underlies most human-caused violence, exploitation, and oppression:

Me and those like me are better and more important than

others. Our feelings, wants, needs, desires, and very lives are worth more than 'theirs.'

This mindset persists in most cultures and reveals itself in manifestations that are both socially sanctioned (i.e. animal consumption, land use, inequitable pay) and non-socially sanctioned (i.e. abuse of companion animals, toxic waste dumping, rape). These two branches of injustice share the same root system; thus one cannot be watered without causing the other to thrive and grow. The outcomes of this mindset include, but are not limited to: patriarchy, racism, sexism, homophobia, heterosexism, classism, genderism, ageism, environmental destruction, speciesism, consumerism, family violence, sexual violence, the prison industrial complex, war... Though seemingly disconnected, these manifestations are connected by the paradigm of perceived superiority. This mindset endures because it has well-established safe havens within the human social norms of most cultural contexts."

Since the global industrial civilization is male-dominated and Western European in origin, its hierarchical system institutionalized the kinds of injustices that are so prevalent today. Therefore, it is no coincidence that four of the richest societies, the United States, Canada, Australia and New Zealand, are the only four countries to have voted against the UN Declaration on the Rights of Indigenous People in 2007[32]. Even as they reluctantly ratified the Declaration after it was overwhelmingly adopted at the UN, these countries passed legislative resolutions stating that the UN Declaration was non-binding on them[33]. These four countries all happen to be colonized countries, majority populated by recent European immigrants, where symbols of overt racism towards their indigenous populations are still prevalent, even in the 21st century. For example, the names and mascots of national sports teams use pejorative terms and symbols for indigenous people, even today, in the United States[34]. It is also no coincidence that these four countries along with Great Britain are the "Five Eyes" nations implementing a total surveillance state, where there is a mere semblance of privacy for the general public, but total secrecy for governments and corporations. As Bruce Schneier, a security expert wrote[35],

"Both government agencies and corporations have cloaked themselves in so much secrecy that it is impossible to verify anything they say; revelation after revelation demonstrates that they've been lying to us regularly and tell the truth only when there's no alternative... All of us are being watched, all the time, and that data is being stored forever. This is what a surveillance state looks like, and it is efficient beyond the wildest dreams of George Orwell."

This is the exact opposite of what should be happening in true democracies, where governments and corporations should be transparent, while the general public should be enjoying freedom and privacy. Instead, corporations are often unwilling to tell us what exactly are in the products they are selling us and what exactly are the chemicals they are injecting underground to extract oil and natural gas. The laws of the land have been clearly orchestrated to support such corporate opacity[36]. The government of the United States has even gone so far as to assert worldwide, extrajudicial, absolute rights, wielded in total secrecy. As Rosa Brooks, the Former Special Coordinator of the US Defense Department Rule of Law and Humanitarian Policy Office, testified before the US Congress recently[37],

"Right now we have the Executive Branch making a claim that it has the right to kill anyone, anywhere on Earth, at any time, for secret reasons based on secret evidence in a secret process undertaken by unidentified officials. That frightens me!"

But there are no signs that governments around the world are about to change any time soon. Their institutional integrity is virtually nonexistent. When governments sign an agreement at the UN that they will halt biodiversity loss by 2020, for example, they act as if that is a green signal to continue destroying biodiversity until 2020. As the climate scientist, Joern Fischer, said recently[38],

"For biodiversity conservation, we are trying at the moment a whole bunch of meetings and setting ambitious targets. The UN set itself a target of no biodiversity loss by 2010 and failed. And it set itself a target of no biodiversity loss by 2020 and it will fail again. And it will fail again because there are simply no strategies in place that would ensure that this target can actually be met. At the moment, we have a lot of talking about things and appealing to things within the same systems that we have used in the past and hoping we will somehow get a different outcome."

This is because world governments are essentially tasked with growing the economy and keeping the hierarchical system stable. That is their main job. The communique of the recently concluded G20 summit began[39],

"Raising global growth to deliver better living standards and quality jobs for people across the world is our highest priority."

It mentioned the word "growth" 29 times in 3 short pages. But if any of the governments were looking to reduce human pressure on the Earth's climate and ecosystems by curbing human population, a recent study published in the Proceedings of the National Academy of Sciences put rest to that notion[40]. It showed that only the ugliest of scenarios, such as a rapidly enforced global one-child policy or the mass die-off of several billion people could alter population trajectories significantly by 2100 to make a dent on our environmental impact in our current course. The PNAS paper concluded,

"That leaves systemic changes to societies' resource use, its forms of energy, its economic structures and its social organization as the crucial moves that can lead to a sustainable civilization."

That is, everything must change. System change is the only path to a sustainable civilization, but such a system change is difficult to accomplish from within the existing reality. Fortunately, as Dr. Vandana Shiva pointed out, the new system is already being built, in bits and pieces. It is based on diversity, freedom, justice, equality, joy, culture and it is springing up through various grassroots social justice movements around the world. Therefore, our task at hand is to coalesce these seemingly disparate social justice movements into a cohesive whole and give it the appropriate structure to create a viable alternative for the current hierarchical socioeconomic system. For all social justice activism is part of the same common struggle for freedom and justice[41]. As Holly Wilson said[42],

"Animal rights, gay rights, human rights - it is all the same battle, fought on different fronts. We are all living beings. We possess a desire to live, love whom we choose, and deserve to live free of brutality and oppression."

While all oppressions are based on the idea that some lives matter less than others, Gordon Allport pointed out many years ago that oppressions of all forms have a common origin[43]. They all begin with words. Words that separate an out-group from an in-group, then lead to denigration of that out-group, to avoidance, to discrimination, then physical attack and ultimately to even genocide and extermination.

Thus all social injustices are built with the exact same bricks in every case, whether it is on the basis of gender, sexuality, disability, or the species of origin. According to David Hufton[44],

"Oppression is the bitter fruit of the tree that is grown, root and branch, from bullying seed."

At its core, bullying is the abuse of power to coerce the weak to act against their will. It is the most fundamental form of violence. In turn, bullying is at the very core of the hierarchical system for no one voluntarily chooses to be at the bottom this system. As Kristof Vanhoutte, Gavin Fairbairn and Melanie Lang wrote[45]:

"Most of us first come across bullying in school, whether as victims, perpetrators or both. But it is much more significant in human affairs than a bit of pushing and name calling in the playground. It is to be found in education at all levels, from kindergarten to university, among both staff and students, in prisons and detention centers, in sport; in politics, both within and between political parties and in workplaces of all kinds. It is found in families, where it manifests itself not only in the squabbling that goes on between siblings, but in domestic violence; in the physical and sexual abuse of children and elders; in the imposition, within some communities, of unwanted marriages, and in the explosions of human emotion that are honor killings. It is found in international trade, with some multi-national companies abusing the power that their financial and business strength gives them to bully suppliers across the globe that provide the products they sell. It is found in the lack of empathy and fellow feeling that leads to the abuse of political power and physical force, by repressive political regimes that suppress dissent through torture and disappearances. It features strongly in the route that dominant groups in some countries and regions of the world have taken in moving from intolerance via discrimination, to genocide."

But above all, bullying is rampant in the Animal Agriculture industry. That's where the "Holocaust on Life" is conducted in earnest within the current socioeconomic system. Animals are maimed, raped, incarcerated and killed by the billions and in terms of sheer numbers, animals constitute more than 99% of the victims of all bullying[46]. We are rightly horrified by the statistic that one billion women living in the world today can expect to be sexually violated within their lifetimes if the current hierarchical system endures. But think of the 70 billion land animals that can expect to be killed on a slaughterhouse floor this year alone! This is why it becomes almost impossible to eliminate social injustices of any kind while continuing the consumption of animal-based products. For how could we possibly eliminate the cultural propagation of gender violence in society, to cite one example, when millions of people are routinely engaged in gender violence on animals in their jobs and come home daily, stressed from that experience? Or when we are constantly consuming the maternal and menstrual secretions of animals, extracted through such gender violence? This is why

Dr. Will Tuttle fingers Animal Agriculture as the root cause of all our social justice ills[47]:

> "Is there an idea that could transform the roots of our culture and create a solid foundation for peace, abundance, and sustainability? I believe there is, and that it has to do with questioning the pervasive influence of animal agriculture... The essential dot connection we are called to make today is between our routine abuse of animals and virtually all of the crises we face, both collectively and individually."

Indeed, sociological researchers have found that someone who believes himself to be superior to animals is more likely to believe himself to be superior to other human beings who are not like him[48]. Conversely, Veganism is a universal salve that heals social wounds of every kind to a large extent and can integrate seemingly disparate social justice activism into a cohesive whole. As the humane educator, Marla Rose, put it[49],

> "I am a feminist. I am vegan for the same reasons that I'm a feminist. It is really as simple as that."

## 7.3 The Greatest Transformation

Veganism is like a four-legged stool with health, ethical, environmental and spiritual reasons for elevating our lives. While the health reasons are becoming increasingly well known with the advocacy work of numerous health and nutrition professionals and organizations, the ethical, environmental and spiritual reasons reinforce them to make a truly compelling case. Thus, Veganism is an essential part of the upcoming fundamental transformation of our global industrial civilization, the greatest transformation in the life of our species. But this time, it is not about concentrating more and more power in the hands of a few, but devolving power to the local level in the hands of the many. This is the metamorphosis and just as in Nature, the Caterpillar has no choice but to become a Butterfly.

Historically, every momentous transformation in human civilization has been accompanied by revolutionary changes in three aspects of human lives[50],

1) In the way we harness energy;
2) In the way we communicate with each other; and
3) In the foods we eat.

About 200,000 years ago, we

1) Discovered the controlled use of fire;
2) Developed spoken language to communicate with each other; and
3) Began eating meat from hunted animals because our controlled use of fire allowed us to cook that meat and made it digestible.

Thus began the dominance of patriarchy as the male hunters assumed more importance than the female gatherers in human societies. The gatherers no longer had to forage over large distances to gather the nutrition needed for human sustenance since the hunters could provide concentrated nutrition in the form of animal flesh. Simultaneously, this strengthened speciesist attitudes within human societies, as animals became objects to be killed for human consumption. Thus sexism and speciesism are the core oppressions from which all other oppressions sprung over time. Hierarchy developed within the patriarchy. The victims of sexist oppression, the women, were partly assuaged when they could oppress other species and feel superior to them.

About 10,000 years ago, during the agricultural revolution, we

1) Harnessed the energy of animals such as cows, buffaloes and horses to plough our fields;
2) Developed writing in order to communicate with each other; and
3) Grew crops of our own liking instead of relying on what Nature provided in the wild.

Instead of humans belonging to Nature, we began acting as if Nature belonged to humans. Not only did we enslave work animals to do our bidding, we enslaved the Earth to produce what we desired. In the resulting agricultural revolution, cities were born where the ruling classes did not do the actual work of raising crops but were fed very well. The social hierarchies developed more layers, resulting in other oppressions such as slavery, classism and casteism.

About 200 years ago, we

1) Began to harness fossil fuels for energy;
2) Developed the printing press for communication, to disseminate information far more efficiently than with just hand written documents; and
3) Repurposed our domesticated work animals to be raised as just food animals.

We developed machines to plough the fields and didn't need the work animals anymore for that purpose, but we continued to enslave them anyway just to milk them and eat them. We developed further layers of hierarchy in our social structures to expand the scope of our human enterprise until it bestrode the whole globe, conquering and colonizing any indigenous civilizations that came in our way. The fossil fuels were to be found in specific locations on Earth and we had to create refining, processing and distribution systems for them. The food animals were most efficiently raised in giant factories as if they were widgets, and then processed into meat packages, refrigerated and distributed to the consumers up and down the social hierarchies. A dominant financial sector arose that siphoned off increasingly larger shares of the wealth, simply as a commission for allocating capital efficiently. Oppressions such as colonialism and racism became much more prominent.

Today, we are poised to undergo yet another transformation, the greatest of them all! This time:

1) We are harnessing solar energy directly and rather than being concentrated in a few locations, it is actually falling on our heads almost everywhere.
2) We are using the Internet to communicate with each other and it

has put the entire accumulated knowledge of all humanity at each and every finger tip.

3) We are transitioning out of animal-based foods to plant-based Vegan foods, which can mostly be grown in local farms without having to rely on large animal husbandry operations with giant processing, refrigeration and distribution systems that are currently spread out over half the globe.

Unlike the previous three major transformations that increasingly concentrated power in the hands of a few and strengthened the social hierarchy, what is occurring today is an entirely radical kind of transformation since all three changes devolve power to the local level, where it becomes easier to implement cooperative and consensual decision-making processes.

This devolution of power is already evident in the US. While the US Congress is quite gridlocked and can barely manage to pass continuing resolutions that maintain the status quo, local governments in cities and municipalities, from Detroit to Seattle to Los Angeles to Tempe, have been promoting urban farming, innovative housing solutions, and other such radical changes. Therefore, the transformation that we're undergoing now is towards a loosely connected global network of densely connected local communities. But, of course, such a revolutionary transformation will need to overcome the resistance of the power elites in the current hierarchical system, who naturally fear the loss of their perceived privileges and the chaos that would occur if the current system collapses.

The fossil fuel industry has been stoutly resisting the growth of the renewable energy sector. But Al Gore, among others, is predicting that a global transition to solar energy will be largely complete by 2030 as solar costs spiral downwards. Prof. Mark Jacobson of Stanford even has a detailed plan of how such a transition might occur[51]. Besides, this plan assumed that the Caterpillar culture will continue unchecked and the energy demands of humanity will continue to soar into 2030! In reality, this is certainly not going to be the case. If the steady state Butterfly economy requires one-third the energy that we use today, which is reasonable considering that the majority of our present activities are unnecessary and wasteful, the transition to solar energy can happen sooner.

The solar energy sector needed just one chink in the armor of the fossil-fuel interests to realize its economies of scale and that chink came in the form of generous German incentives. When combined with the declining costs of electrical storage batteries, thanks to the success of Tesla Motors, gasoline engines will also likely die a slow death before 2030. Therefore, fossil fuels will literally become dinosaur fuels by 2030. Local collection of solar energy, coupled with the local energy storage using batteries will mean that the large utility companies of the past would become obsolete as well. In fact, in Europe, the top 20 utility companies lost over 60% of their stock market value between 2010 and 2015[52], just when the DAX stock market index gained 60%. What's happening in Europe will inevitably happen worldwide!

In the communications arena, the power elites have also been trying to subvert the Internet, instituting an elaborate surveillance infrastructure so that every electronic transmission of every individual can be stored, accessed and searched, for all time, ostensibly to keep us all "safe from terrorism". Then Edward Snowden came along and spilled their secrets[53]. He showed that the watchers had an expansive definition of "terrorism". If you were kind to animals, you were a "terrorist," because you were a threat to the hierarchical system. If you were kind to the environment, you were an "eco-terrorist," for the same reason. All passive bystanders were treated as "potential future terrorists" within the current system. Therefore, all of us were being watched, all the time!

But Edward Snowden also revealed that if Internet communications were encrypted with reasonably strong cryptographic codes, then the watchers became blind. They and the thousands of math Ph.D.s who work for them, don't know how to break these codes, most likely because these codes are truly unbreakable. Therefore, he showed that if we routinely encrypt all our communications and develop email and social network applications to do that seamlessly for the casual user, then the Internet will become what it was always intended to be: a communications technology that frees us all from the clutches of a few. We can then have true privacy for the individual and demand absolute transparency from our institutions, which is the only way we can assure an open government of the people, by the people and for the people. In this post-Snowden era, various open source software consortia are already implementing such seamless encryption systems.

## 7.4 The Vegan Metamorphosis

The final change that will help dismantle the hierarchical system of the Caterpillar is when we transition to agro-ecologically grown, local, plant-based foods. To grow animal-based foods, humans had to create vast monocultures and an industrial infrastructure encompassing almost half the land area of the planet. Compared to the standard American diet, a strictly plant-based diet requires 18 times less land to grow! This is why the most effective act of rebellion against the hierarchical system these days is to grow your own produce and to go vegan. This change to a primarily local, plant-based food system needs to occur for the emergence of the Butterfly within the next 1-2 decades as the non-hierarchical solar energy system and the non-hierarchical, encrypted Internet communication systems both fall into place, also during this same time span.

At first glance, it may appear that the odds are stacked against this behavioral change in our food habits that needs to occur so quickly. Indeed, when we consider the historical behavioral trends in the case of tobacco smoking, we might get discouraged. The anti-tobacco campaign was initiated in the US, top-down, at the behest of the Surgeon General in 1965[54]. Despite the immense persuasive powers of world governments leading the anti-smoking campaign, the total world consumption of tobacco had not yet reached its peak in 2015, 50 years later[55]. The US Surgeon General's office was very emphatic in that campaign, insisting that people quit smoking entirely, explicitly guilt-tripping consumers on the second hand effects of smoking and even going so far as to run advertisements telling consumers that,

"Smoking Kills."

In our case, we need to achieve not just a peaking but also a substantial reduction in the world consumption of animal foods within the next 1-2 decades for the Butterfly to emerge. For this to occur, we either have to institute a strict top-down rationing of animal foods worldwide or a lot of us have to voluntarily reduce our consumption of animal foods to ZERO, i.e., go vegan. Leave aside top-down rationing, we can't even expect the same level of enthusiasm from the US Surgeon General's office or any branch of world governments in an anti-meat/dairy/fish/eggs campaign as in the anti-smoking campaign. If anything, it will be the opposite as most governments currently use taxpayer funds to subsidize and promote the consumption of animal-based foods. Derivative animal products are to be found everywhere in our consumer products, from the glue in our cars to the leather in our shoes to the additives in our processed foods[56]. Government food programs also indirectly subsidize these derivative animal products. Therefore, it would be tremendously disruptive to the current socioeconomic system if the main reason for animal husbandry disappears when people stop consuming animal-based foods. The oppression of animals and by extension, the oppression of indigenous people, is at the very base of the world economic engine that is keeping the current hierarchical system going. Therefore, the power elites in the hierarchical system will work to prevent this oppression from being dismantled since they will view its continuation as necessary for the continuation of their power.

But there are some key elements of weaknesses and contradictions in the current system that can help this transition happen quickly. In any long-standing systematic oppression, whether it is the oppression of colored people, LGBT people, animals or the destruction of the environment, there are usually 4 groups of participants:

1) The perpetrators;
2) The victims;
3) The onlookers; and
4) The moderates.

In general, it is the fourth group of participants, the moderates, who are critical in keeping that oppression going even when a majority of the onlookers want the oppression to stop. The moderates are typically widely trusted sources, who appear to be on the side of the victims and the majority onlookers, but they are indirectly serving to keep the oppressive system stable by enabling the perpetrators. But the moderates are also the most vulnerable part of the systematic oppression, for they are the ones who are caught in a cognitive dissonance and therefore, amenable to public persuasion. For example, in the case of the legalized oppression of LGBT people in the US, many members of the Democratic party played the role of the moderates and it was only by pressuring them and not the die-hard Republicans that the LGBT rights movement gained significant legislative victories[57]. In retrospect, it is only by raising the awareness of the "White Moderates" in his letter from a Birmingham Jail that Rev. Martin Luther King, Jr., made his breakthrough in the civil rights movement and there is a lesson in this for all of us[58]:

"I must make two honest confessions to you, my Christian and Jewish brothers. First, I must confess that over the past few years I have been gravely disappointed with the white moderate. I have almost reached the regrettable conclusion that the Negro's great stumbling block in his stride toward freedom is not the White Citizen's Counciler or the Ku Klux Klanner, but the white moderate, who is more devoted to "order" than to justice; who prefers a negative peace which is the absence of tension to a positive peace which is the presence of justice; who constantly says: "I agree with you in the goal you seek, but I cannot agree with your methods of direct action"; who paternalistically believes he can set the timetable for another man's freedom; who lives by a mythical concept of time and who constantly advises the Negro to wait for a "more convenient season." Shallow understanding from people of good will is more frustrating than absolute misunderstanding from people of ill will. Lukewarm acceptance is much more bewildering than outright rejection."

When the White Moderates, stung by this public exposure of their role in perpetuating racism, showed up in large numbers at the Washington Mall and supported the Civil Rights Movement wholeheartedly, it gained an unstoppable momentum.

In the oppression of animals, there are four kinds of moderates:

1) The faith moderates;
2) The ethical moderates;
3) The environmental moderates; and
4) The health moderates.

The faith moderates are perhaps the most important class of moderates as they have the potential to change very quickly and have tremendous grassroots organization capabilities. Most are respected members of the community who say the right things and have their hearts in the right place regarding the oppression of animals, but they contradict their words with their actions, thereby giving the green light to their followers to do the same. For instance, Pope Francis wrote in the *Laudato Si*[59],

> "It is contrary to human dignity to cause animals to suffer or die needlessly."

But despite these words, Pope Francis has continued to consume animal foods. For instance, he was reported to have eaten a veal and lobster dinner during his visit to New York City[60]. This made it seem as if it is necessary to eat animal foods for human wellbeing even though the American Dietetic Association has clearly said that it is unnecessary to eat animal foods of any kind at any stage of our life cycle[61]. Just as Pope Francis did wonders for the acceptance of climate science in the *Laudato Si*, he now has the power to step forward and do the same for the acceptance of nutrition science and help save humanity from planetary scale catastrophes.

In the past, animal foods were consumed mainly on special occasions and religious customs were in place to ensure that these special foods were shared with all members of society, including the poor. During those days, people relied entirely on locally grown foods for sustenance. It was probably necessary to consume the flesh and secretion of animals that metabolized most of the vegetation in the local area so that humans could get all the essential nutrients for their sustenance. Otherwise, the vegetables, fruits and grains alone may have left people deficient in certain nutrients. In fact, during his speech to the London Vegetarians Union in 1932, Mahatma Gandhi lamented that he tried going vegan several times and failed because of health reasons[62]. But in modern times, we have the technology to grow almost any plant-based food anywhere and we have access to an abundance of non-local plant-based foods in our supermarkets. We can synthesize plant-based equivalents for almost any animal food. This is why it is perfectly feasible to lead a vegan lifestyle especially in the global North and there is absolutely no need to eat animal foods of any kind at any time. Besides, it is particularly unhealthy to gorge on animal foods as we have been habituated to do these days, since the planet is marinating in toxic effluents from all our industrial activities. Those toxins are working their way up the food chain, increasing their concentration by orders of magnitude at every step.

The second class of moderates, the ethical moderates, are typically Animal Welfare advocates, who can be found decrying the oppression of the animals, while they negotiate with the perpetrators to "reduce the suffering" of the animals. They unconsciously play the "good cop" role vs. the perpetrators "bad cop," but the net result is that they prolong the oppression well beyond the time when a majority of the onlookers have stopped tolerating it. For instance, such moderates can be found in Animal Welfare organizations that certify various levels of "humane" treatments for slaughtered animals.

A unique aspect in the oppression of food animals is that most of the onlookers are also indirect oppressors since they are paying the perpetrators to do the oppression for them. But when such onlookers feel that the oppression is going too far, it is such ethical moderates who step in to assuage the guilt of the onlookers, telling them comforting stories about "humane meat," "grass-fed beef," "cage-free eggs," etc. Or they tell distracting stories about other cultures that deflect attention from the oppression that's occurring right under the noses of the onlookers. Since such distracting stories raise donations from outraged onlookers for the ethical moderates, this becomes a self-perpetuating action. For example, TIME magazine had this blaring headline recently[63]:

"The World's Largest Animal Slaughter Festival has begun in Nepal!"

The story had to do with a local festival that occurs once every five years in Nepal. It was about some obscure, ancient, Nepalese festival, involving a miracle that someone experienced, following which he sacrificed a water buffalo and donated the meat to the hungry[64]. The practice has now been continued for centuries, with the result that thousands of buffaloes are sacrificed in this "Gadhimai" festival once every five years in Nepal. This blaring headline in TIME magazine got animal rights activists in the US outraged over the "barbaric practices" of Hindus in Nepal, while distracting them from the slaughter of 60 million turkeys happening locally in the US for Thanksgiving, at exactly the same time. Clearly, the magnitude of the Thanksgiving slaughter dwarfed the "Largest Animal Slaughter" festival of Nepal, but that fact got lost in the outrage. Such distractions serve to perpetuate both atrocities and thereby maintain the status quo.

Firstly, it is easy to get activists in one culture riled up over seemingly strange atrocities in another culture. It is always much easier to bash "others" than to do soul searching at home. Meanwhile, the Nepalese will naturally resent foreigners judging them over the barbarity of their animal sacrifices in the thousands, especially when those same foreigners are murdering turkeys in the millions while giving thanks on a single day. When such resentment is bred, the practice will gain support in Nepal out of defiance. The activists in the U.S. raised substantial funds to spend their energies focusing on the Nepalese atrocity instead of focusing on their local atrocity. The cycle continues.

In the same way, Sea Shepherd's focus on Japanese whaling perpetuates both Japanese whale hunts as well as the incidental killing of thousands of whales in drift nets during Western industrial fishing[65]. The Japanese resent that Capt. Paul Watson of Sea Shepherd appears to be focusing on their "relatively minor" whaling atrocity, when compared to the industrial carnage in commercial fishing. They prefer that Capt. Watson tell American and European fishing fleets to clean up their acts first.

Now, if only a Japanese national had been leading Sea Shepherd, then the Japanese might heed his or her words better. That is, even though Capt. Watson is undoubtedly sincere about trying to protect all whales, he serves as an ethical moderate when he focuses on Japanese whale hunts, since there are many intersecting oppressions, racism, sexism, speciesism, etc., of which speciesism is just one. But by raising the consciousness of such moderates, it is possible to make headway, for most of them truly want to be effective.

With respect to raising the awareness of environmental moderates, Kip Andersen and Keegan Kuhn's documentary, *Cowspiracy: The Sustainability Secret*, is destined to become a cultural phenomenon in the annals of environmental history[66]. After watching *Cowspiracy*, Chris Hedges, the noted author and social justice activist, and his family, decided to go vegan recently[67]:

> "My attitude toward becoming a vegan was similar to Augustine's attitude toward becoming celibate—"God grant me abstinence, but not yet." But with animal agriculture as the leading cause of species extinction, water pollution, ocean dead zones and habitat destruction, and with the death spiral of the ecosystem ever more pronounced, becoming vegan is the most important and direct change we can immediately make to save the planet and its species. It is one that my wife—who was the engine behind our family's shift—and I have made...
>
> We have only a few years left, at best, to make radical changes to save ourselves from ecological meltdown. A person who is vegan will save 1,100 gallons of water, 20 pounds CO2 equivalent, 30 square feet of forested land, 45 pounds of grain

and one sentient animal's life every day. We do not, given what lies ahead of us, have any other option."

The facts about the environmental impacts of Animal Agriculture that *Cowspiracy* presents are nothing new. But facts alone don't sway people. The way *Cowspiracy* presents the facts, by first shedding light on the world's big environmental organizations grips the attention of the audience. We realize that the environmental organizations appear to be deliberately ignoring the impact of Animal Agriculture on the environment, thereby facilitating the destruction of all the world's forests and all the fishes in the ocean within the next two to three decades, contrary to their stated missions. Catching these organizations in their evasions and half-truths and then finally getting them to admit that Animal Agriculture is indeed the primary cause for the environmental ills on the planet, sets the stage for the later part of the movie as the audience becomes more receptive to the facts being presented to them. Otherwise, the facts alone could have been conveniently ignored as "Vegan propaganda" by an audience that's naturally averse to being told that they are doing horrible things to the environment every day through their daily habits. For instance, it was important for *Cowspiracy* to get the California Water Resources Board (CWRB) officials to admit that Animal Husbandry had, by far, the largest water footprint among all human activities. Then the audience becomes more receptive to the fact that every hamburger requires 660 gallons of water to produce, especially since the CWRB officials don't dispute this figure in the movie. Finally, when the movie is finished, we realize that there is no way to consume animal-based foods of any kind without adversely impacting the environment.

Fortunately, the public exposure of these inconsistencies is a tremendous motivator for change. As Johann Hari documented in his excellent investigative article in the Nation magazine, it is only those environmental organizations that accepted corporate funding in the 80s that grew large and became the "Big Green" organizations of today[68]. They accepted funding from the big polluting corporations in exchange for some green credentials bestowed on the corporations and chose to look the other way when it comes to issues such as Animal Agriculture. As a consequence, millions of acres of forests are being destroyed each and every year with little or no protest from the Big Green organizations, with all the attendant hardships towards indigenous communities and wildlife, not to mention the destruction of the planet's biological treasures, those still-intact ecosystems.

In this post-*Cowspiracy* era, the Big Green organizations are being confronted in the social media when they do any business-as-usual promotions of animal foods these days. When Al Gore's Climate Reality Project partnered with Ben and Jerry's ice creams to support a "I'm Too Hot" campaign to "serve up climate truth," the strong reactions in the social media happened just one month after *Cowspiracy* premiered[69]! Since then, the awareness of such linkages has been growing exponentially. Such exposure of the environmental moderates strengthens the Vegan movement from the ground up.

Kip Andersen and Keegan Kuhn are planning to release a follow-up documentary to *Cowspiracy* called *"What the Health*[70]," which shows how health-care organizations such as the American Diabetes Association are failing to inform the public about the health impacts of animal-based foods. These organizations are the "health moderates" that round up our moderate list.

Moderation is never a valid response to a moral issue. We don't pledge to refrain from gay-bashing one day a week. We don't pledge to refrain from lynching black people before 6pm each day. Therefore, moderate steps such as the pledge to go Meatless on Mondays[71] or Mark Bittman's pledge to go "Vegan Before 6pm (VB6)[72]" are appropriate only as intermediate steps on the road to eventual Veganism. They are inappropriate long-term responses to the suffering of animals, the suffering of indigenous communities and the destruction of the environment, not to mention our own ill health. For why would we want to ruin our health, torture animals, oppress indigenous people and destroy the planet on six days each week or after 6pm each day? As long as the forests of the world are being destroyed and the ocean is continuing to be overfished, every incremental morsel of animal-based foods consumed anywhere should be rightly seen as a bad idea whose time has past.

Moderation is never a valid response to addictions either. This is why there are no organized smoking moderation programs, but only smoking cessation programs. From a neuro-chemical point of view, animal foods like meat, fish, eggs and cheese are indeed addictive substances. When we consume these substances habitually, we rapidly and reproducibly alter the bacterial composition of our guts[73]. But unlike alcohol, tobacco or drug addictions where the victims are ruining their own health and perhaps the well being of those around them, our addiction to animal foods has planetary scale consequences, not just personal health consequences. The worst part is that most of us were involuntarily enticed into this addiction when we were children and deserve to be treated with compassion and with the necessary support to help overcome it.

This is why the faith community is key to making the Vegan transition happen quickly. When the faith community responds affirmatively and shoulders this responsibility as the planet requires, then it would make religion more relevant to the youth of the world as well. The Hindu faith community has taken the first step already in the 2015 Hindu Declaration on Climate Change. This is the first major religion declaration that has explicitly called for the worldwide adoption of a plant-based vegan diet. Unlike the vegan transition at the grassroots, which is being led by the younger generation - the Miglets, it is truly comforting to see one of the oldest wisdom traditions in the world take the lead on this issue within the faith community. For as the Canadian journalist, playwright and novelist, David Macfarlane wrote [74]:

""Because we all liked cheeseburgers so much" is going to sound pretty stupid when humankind is hauled into the principal's office and asked to explain how the planet got destroyed."

Perish the thought of ever having to deliver such an explanation. In our story, the Principal had always been in charge! This power of

Love is so strong that despite the blatant obfuscations of governments, the environmental organizations and the media and despite the extremely addictive qualities of the animal foods in question, millions of people have been adopting this compassionate lifestyle! The metamorphosis is exponentially increasing in its rapidity and intensity, year after year.

## 7.5 Gandhi in the 21st Century

The Vegan movement has the same role to play in dismantling the hierarchical system of the Caterpillar in the early 21st century as the "Khadi" movement, spearheaded by Mahatma Gandhi, played in dismantling the British colonial empire of the early 20th century[75]. Gandhi fueled the Khadi movement in the early 20th century by writing magazine articles and just plain people-to-people persuasion, without the Internet, without cell-phones and without social media. It is very instructive to study how he got hundreds of millions of people to take voluntary action at once.

The year was 1914. Gandhi, a middle-aged lawyer, dressed in a finely tailored British suit and tie, embarked on a ship to travel from South Africa to India with a brief stopover in England. He was an accidental activist, thrust into that role when he was thrown off the first-class compartment of a train in South Africa, for being colored[76]. But now he was sailing out to join and possibly spearhead the Indian independence movement, to take on the mightiest Empire that the world had ever seen until that time.

Gandhi was received as a hero in India since he had been instrumental in raising the morale of the Indian people through his 1909 monograph, *Hind Swaraj*. After visiting the villages, towns and cities of India over the next three years, Gandhi announced his grand idea for taking on the British empire: Indians must continue with their ongoing agitations, but in addition, they must change their clothes from British clothes manufactured in the mills of Manchester to simple, homespun "Khadi" clothes.

At first, Gandhi's plan was met with some ridicule in Indian intellectual circles as magazine articles from those times show[77]. How could the bitterly divided people of India take on the mightiest empire the world had ever seen by changing their clothes? But he had the backing of a few key intellectuals, including Rabindranath Tagore, a poet Nobel Laureate, who saw the wisdom of concerted action, which could unite the people of India. But more importantly, undermining the British textile industry would severely impact the economic might of the British empire[78]. At that time, the textile industry was one of the largest industries in Great Britain. The Khadi movement was simple to join, had a substantial impact and united the people of India in a common, spiritual bond.

The Khadi movement was born, in 1918. At first, the colonial rulers ignored the Khadi movement, even as Gandhi waged a tireless, grassroots campaign. Gandhi wrote in the Navjivan magazine in 1925[79],

"It is my duty to induce people, by every honest means, to wear Khadi."

Clearly, Gandhi wasn't satisfied with moderate half steps, since he was interested in turning people into passionate activists on behalf of the Khadi movement and, thus, the Indian Independence movement. This was key to the rapid spread of the movement. Much to the consternation of the colonial rulers, within a dozen years after it was founded, by 1931, the Khadi movement had managed to bankrupt the textile mills of Manchester[80], paving the way for the eventual independence of India sixteen years later, in 1947. Gandhi was a genius for framing the Indian freedom struggle, not as one between two countries, but as one between the working class on the one hand, and the ruling power elites on the other. He rightly observed that the textile mill workers of Manchester were also the oppressed victims of industrialization and that it was the English East India Company and not the people of Great Britain that began the colonization of India. Gandhi was also a genius for recognizing that it is only personal changes executed in concert that can lead to a social transformation. The

Khadi movement truly united the people of India in a common spiritual bond, helping them tide over their other vast differences.

Now, almost a century later, we are faced with a very similar situation, but on a global scale. Gandhi was fighting for the independence of India from colonial rule. We are now fighting for the transformation of our socioeconomic system, to ensure the survival of our children and grandchildren and, indeed, to ensure the survival of all life on Earth as we know it. The British Rulers, who clung to the old idea of a vast British colonial empire, which the sun never set on, opposed Gandhi. Likewise, we can expect to be opposed by those who will cling to the old idea of a hierarchical system with rank privileges. Gandhi inspired the people of India to make that one simple change, to take that voluntary step of changing their clothes. We need to inspire people in industrial societies - specifically, all those who have access to food abundance - to take that voluntary step of changing what we eat, to go vegan. For the Animal Agriculture industry of the early 21st century is the global equivalent of the textile industry of the early 20th century in Great Britain. It is one of the largest industries in the world with, by far, the largest footprint on Nature.

Thus the Vegan movement, without a doubt, has the same potential for personal and social transformation in the 21st century globally, just as the Khadi Movement did in 20th century India. Just like the Khadi movement, it is simple to join, it has a substantial impact and it has spiritual connotations. It can truly unite the people of the world in a common bond, helping us tide over our other vast differences. At the moment, at an individual level, it is a far more practical step than foregoing the use of fossil fuels in our daily lives. Is it any wonder that countries like Germany have seen an 800% increase in vegans in just three years by 2013[81]? It is thrilling that the Miglets in the affluent world are leading this transformation!

### 7.6 The AhimsaCoin Economy

Just as the Khadi movement was a necessary first step on the road to Indian independence in the 20th century, so is the Vegan movement also just a necessary first step on the road to global sustainability in the 21st century. Veganism only addresses the cultural propagation of systemic violence, but not the structural aspects of such violence. It is not just the types of foods, clothes and shoes we consume that need to change, but our entire consumer culture needs to change on our road to sustainability. As the economist, Charles Eisenstein, has pointed out, when we dig deeper into the ascendance of the consumer culture in our industrial societies, we see the central role of money, specifically our debt-based monetary system and the competition for profits, at the root of it[82]. It is the centrally controlled, debt-based monetary system that necessitates a hierarchical organization of our socioeconomic system. In our current system, we issue currency for debt, but never issue currency for interest on the debt, with the result that borrowers have to compete with each other to acquire the currency for paying back the interest on the debts. Alternately, the economy has to continuously grow in order to grease the flow of currency back to the lenders who truly occupy the top positions on the totem pole. There is an apocryphal saying attributed to the House of Rothschilds[83],

"Give me control of a nation's money and I care not who makes its laws."

Such is the concentration of power afforded by the central control of money. Over time, the Rothschild progeny and a few others have indeed managed to wrest control of the world's monetary systems. The Georgetown historian, Carroll Quigley, described their plan in 1966[84],

"The powers of financial capitalism had a far reaching plan, nothing less than to create a world system of financial control in private hands able to dominate the political system of each country and the economy of the world as a whole... Their secret is that they have annexed from governments, monarchies, and republics the power to create the world's money."

As the economist, Thomas Greco, Jr., explained[85],

"The entire machinery of money and banking has been contrived to centralize power and concentrate wealth... Money is undemocratic as it concentrates power in the hands of a few unelected people who are unresponsive to the needs of the people."

Such centralized control of the monetary system was perfect for the Caterpillar phase as it led to the rapid development of the world's military capabilities over the past century. But it is reaching the end of its usefulness. Now, some respected thinkers have claimed that we can address the ills in the hierarchical socioeconomic system by wresting broad democratic control over private capital, through government intervention. The journalist, syndicated columnist and author, Naomi Klein, described what such an intervention would entail[86]:

"Responding to climate change requires that we break every rule in the free-market playbook and that we do so with great urgency. We will need to rebuild the public sphere, reverse privatizations, relocalize large parts of economies, scale back overconsumption, bring back long-term planning, heavily regulate and tax corporations, maybe even nationalize some of them, cut military spending and recognize our debts to the global South. Of course, none of this has a hope in hell of happening unless it is accompanied by a massive, broad-based effort to radically reduce the influence that corporations have over the political process. That means, at a minimum, publicly funded elections and stripping corporations of their status as "people" under the law. In short, climate change supercharges the pre-existing case for virtually every progressive demand on the books, binding them into a coherent agenda based on a clear scientific imperative."

Russell Brand, the comedian and philosopher had a similar prescription as well[87]:

"A socialist egalitarian system based on the massive redistribution of wealth, heavy taxation of corporations and massive responsibility for energy companies and any companies exploiting the environment. I think the very concept of profit should be hugely reduced. David Cameron says profit isn't a dirty word. I say profit is a filthy word, because wherever there is profit, there is also deficit."

Naturally, conservatives are truly repelled by this prospect of a "new world order" and "government takeover" of the socioeconomic system, leading to the current impasse[88]. Besides, such a response still leaves a central, debt-based monetary system intact, which fundamentally fosters a hierarchical socioeconomic system, dependent on competition and economic growth. As long as our socioeconomic system is hierarchical, we would need selection mechanisms to determine who gets to be on top and who languishes at the bottom, which fosters violence, injustice and inequality. Even if we manage to eliminate all vestiges of racism, sexism, heterosexism, speciesism, ableism, ageism, etc., and restore historical inequities for all oppressed groups, we would still need to select for privileges, perhaps on the basis of merit and competition. That would still result in a system of artificial scarcity, wage slavery, debt slavery and structural violence, just to name a few nasty by-products of a centrally controlled, debt-based monetary system.

When leading thinkers like Naomi Klein argue that capitalism is the problem, they automatically seem to assume that socialism is the solution. But socialism failed badly in the Soviet Union. If the option is between government owned means of production as opposed to the private owned means of production, then the evidence accumulated over the past two hundred years shows that the private control is actually more efficient at allocating the planet's precious resources. Therefore, capitalism is preferable to central planning and rank socialism, but that capitalism needs to be "saved" once again.

Imagine that we can devise an alternate, distributed monetary system, in which basic equality and environmental sustainability are designed in, from the ground up. After all, why should money be centrally sourced? Money is just an accounting tool for trading things of value and those things of value are distributed among all of us to begin with. As the economist Mike Maloney described it[89],

> "Your true wealth is your time and freedom. Money is just a tool for trading your time. It is a container to store your economic energy until you're ready to deploy it."

Therefore, it doesn't make any sense for money to be centrally sourced, especially in this Bitcoin blockchain era[90]. Bitcoin is a digital payment system invented by Satoshi Nakamoto, who published the invention in 2008 and released it as open-source software in 2009. The system is peer-to-peer and transactions take place between users directly, without a trusted intermediary, such as a Bank. The unit of currency in the system is a "bitcoin," which is a virtual currency that can be exchanged for money in the current socioeconomic system at prevailing exchange rates, if need be.

The blockchain technology underlying Bitcoin has demonstrated that banking can be accomplished reliably in a distributed fashion, using the power of millions of computers on the Internet to achieve the same level of trust, if not greater trust, than that provided by large, centralized banks. While the Bitcoin implementation of the blockchain technology uses a central allocation of currency and is not concerned with equality, it is easy to modify the Bitcoin implementation to accomplish our twin objectives of environmental sustainability and basic equality. Money is also a demand on the Earth's resources and therefore, we can assure environmental sustainability if our alternate currency is issued in the form of ecological credits, measured in terms of a global ecological footprint. Producers who use natural resources would be required to retire an appropriate number of ecological credits in order to ensure that human impact on the planet never exceeds a suitable fraction of the biological capacity of the planet. Prof. E. O. Wilson, has recommended that at least half the land area of the planet should be reserved for wildlife and natural ecosystems and these wildlife reserves should be contiguous from North to South to allow for ecosystem migration as the Earth's climate changes in the future[91]. It's easy to build in such a constraint on the issuance of ecological credits so that our total human impact on the planet never exceeds acceptable limits. It's also easy to close the loop in such a system since we have satellite technology already deployed that can verify compliance of the human footprint on the planet with the accounting in the system.

In addition, since our objective is to transition to a sustainable, nonviolent society, it is important to build in a measure of basic equality in our alternate currency since equality is closely related to the structural violence in society. As James Gilligan, the former director of the Center for the Study of Violence at Harvard Medical School said[92],

> "If there is one principle I could emphasize that is the most important principle underlying the prevention of violence, it would be equality. The single most significant factor that influences violence is the degree of equality vs the degree of inequality in that society."

If people are treated equally by the monetary system from the outset, then that mitigates poverty at its roots. As Gandhi rightly observed, poverty is the worst form of violence. Besides, the pyschosocial stress of inequality has public health consequences. People get sick when they feel poor, even if they are materially richer than some of the richest people who lived just a century ago[93].

One such distributed monetary system can be conceived in the form of "AhimsaCoin". AhimsaCoin would be similar to Bitcoin, but with authenticated members automatically receiving one AhimsaCoin every 50 minutes into their account during their lifetime. The authentication could be any form of unique identification technology that minimizes the probability of ghosts sucking up the ecological footprint on behalf of unscrupulous users. Think of AhimsaCoins as the space travel rewards bestowed upon each human passenger on spaceship Earth, which is then traded for real goods and services.

Each AhimsaCoin entitles the owner to the productivity of 1 $m^2$ of the Earth's surface for one year so that humanity's global ecological footprint does not exceed half the Earth[94]. As the human population on Earth increases, the frequency of AhimsaCoin issuance will decrease to 1 every 51 minutes, 1 every 52 minutes, and so on, so that there is feedback built into the system to encourage human population stabilization. Also, the frequency of AhimsaCoin issuance should, perhaps, never increase over time so that there would never be an incentive for drastic population reduction either. AhimsaCoins would be retired by producers who extract resources from the Earth, as if they were ecological credits. Private enterprises would raise capital by issuing stock and collecting a sufficient number of AhimsaCoins for their stated purpose.

Capitalism in the AhimsaCoin system would work in much the same way, as we know now, but without credits. Since the total footprint of human activities on Earth would be constrained, economic development would occur only through true productivity improvements and innovation. Besides, since every human being would have ready access to AhimsaCoins sufficient to meet his/her basic needs, it would be impossible for exploitative enterprises to flourish in such an economy. People would do something only if they are truly inspired by the work, while routine work would need to be automated by the private enterprises. Indeed, corporations that inspire workers in the AhimsaCoin economy would be transparent, open-source, cooperative enterprises. This would naturally foster a gift economy where volunteerism flourishes. Those who earn extra AhimsaCoins by serving their fellow humans voluntarily would truly be celebrated. Those who hoard AhimsaCoins would be viewed as heroes since they would actually be reducing the overall human footprint on the planet! Cradle-to-cradle enterprises would have an advantage, as they would need less AhimsaCoin capital to operate. Finally, a distributed monetary system like AhimsaCoin would be perfectly matched with the distributed energy, communications and food systems of the future. The fact that AhimsaCoin codifies basic human equality and thus dignity, along with environmental sustainability, right from the outset, would just be an added bonus!

We are at a socioeconomic crossroads that reminds me of the dilemma facing the 90s era Internet community as we were deciding between a hierarchical communications system architecture for the Internet (ATM) vs. a non-hierarchical, distributed architecture (Ethernet)[95]. Despite grave misgivings expressed by the telecommunications specialists of those days that a distributed, non-hierarchical architecture would be unreliable, the Internet has turned out to be remarkably robust and stable even as it grew exponentially. Perhaps, our Internet experience would serve as a springboard for the adoption of a non-hierarchical, distributed currency system like AhimsaCoin, as we look towards our Butterfly future.

## 7.7 The Half-Earth Solution

If it's true that the melting of the West Antarctic Ice Sheet is unstoppable and our major coastal cities drown, humans can rebuild cities inland. If it's true that the scorching of the American Southwest will be irreversible, humans can move out of Nevada into Utah. If it's true that it is going to be impossible to grow food in Syria, humans can migrate from Syria into Europe.

We humans have the distinct ability to adapt ourselves quickly to any environment on Earth. Other species don't. They don't wear clothes, they don't have air-conditioners, they don't have heaters and when they choose to move to a suitable habitat, they encounter man-made obstacles such as roads and cities that impede them. As a result, many wild animals have already gone extinct. But there is still much that we can do to restore the integrity of ecosystems and help Life flourish on this planet.

A few years ago, Prof. E. O. Wilson enunciated "Wilson's law"[96]:

"If we save the living environment, the biodiversity that we have left today, we will also automatically save the physical environment. If we only save the physical environment, then we will ultimately lose both."

That law has been emblazoned on our Climate Healers web site ever since. From that law, we derived our core guiding principle:

**"Compassion for all Creation is Infinitely Sustainable!"**

Prof. E. O. Wilson just published his 32nd book entitled, *Half-Earth: Our Planet's Fight for Life*[97], which makes an eloquent case for half the Earth to be set aside as a permanent preserve for the benefit of the 20-100 million other species on the planet so that they have a chance to recover from the depredations of our Caterpillar phase. Prof. E. O. Wilson has asked that this Half-Earth preserve should be in the form of contiguous North-South corridors so that species can migrate freely in response to changing climactic conditions. Human passages through these corridors should be designed with respect for the well being of wildlife. Just think of highway crossings on pillars within these wildlife corridors, instead of our current practice, expecting wild animals to cross our highways at their peril.

In a recent interview with the New York Times, Prof. E. O. Wilson sounded a note of optimism[98]:

> "Large parts of Nature are still intact — the Amazon region, the Congo Basin, New Guinea. There are also patches of the industrialized world where Nature could be restored and strung together to create corridors for wildlife. In the oceans, we need to stop fishing in the open sea and let life there recover. If we halted those fisheries, marine life would increase rapidly. The oceans are part of that 50 percent.

> Now, this proposal does not mean moving anybody out. It means creating something equivalent to the U.N.'s World Heritage sites that could be regarded as the priceless assets of humanity. That's why I've made so bold a step as to offer this maxim: Do no further harm to the rest of life. If we can agree on that, everything else will follow. It's actually going to be a lot easier than people think."

In a world that is rapidly transitioning to veganism, this bold Half-Earth proposal is feasible to implement. In that vegan world, most of the 45% of the land area of the Earth that is currently being used for Animal Agriculture will be freed for "re-wilding" [99]. In a world where equality is structurally guaranteed, surely there would be plenty of volunteers to help shepherd this re-wilding! Just imagine 60 million SAI sanctuaries spread throughout the Earth and you will have a vision of the Eden that Prof. E. O. Wilson envisions. Such an Eden can and must be regenerated!

Prof. E. O. Wilson estimates that if the Half-Earth solution is implemented, then we can preserve 85% of the species extant today and limit the long-term extinction to about 15%. That would make this Sixth Great Mass Extinction event[100] to be the least damaging compared to the other five Great Mass Extinction events in the Earth's history.

# 8. The Mewar Angithi

*"Good judgement comes from experience, but experience comes from bad judgement"* - Mullah Nasruddin.

I have spent the last eight years working with indigenous communities in India on our global environmental crises. These communities experience climate change, biodiversity loss, desertification and toxic pollution first-hand and they have been among my best teachers. During my interactions with them, I have been amazed at the depth and breadth of knowledge that they possess on the biodiversity in their forest habitats. It is my dearest wish to make their life easier so that they can continue to stay in their forest homes, if they so choose. In the Caterpillar culture, such indigenous communities are being forced to move into cities to seek employment as bricklayers or manual laborers as their forest homes get destroyed. Instead, we would all be much better off if these communities can stay in their forest homes and help pilot the re-wilding of the planet in the Half-Earth solution!

Their forest homes are dying mainly because of livestock production and fuelwood extraction. Livestock production can only be addressed as the demand for livestock products subsides, but that requires behavioral changes among the affluent communities of the world. Fuelwood extraction, on the other hand, is a local issue as far as these communities are concerned. Fuelwood burning is the second most significant quantitative reason for deforestation worldwide, behind only livestock production[1]. The soot from incomplete combustion of fuelwood is also a potent greenhouse gas. As it gets deposited on the ice in the Arctic or in the Himalayas through wind currents in the Northern hemisphere, the black soot absorbs solar energy whereas the original ice would have reflected it. As a result, the Arctic region and the Himalayan third pole are heating up faster than almost every other part of the Earth due to climate change. Since the industrial chemicals that we have emitted into the atmosphere come down in the rain, get absorbed by trees and get embedded in the trunks and branches of trees worldwide, fuelwood smoke has also become a potent human health hazard, contributing to a loss of as much as eight years in life span for the women who cook with wood[2]. That is the level of lifespan reduction that can be expected from smoking the equivalent of two packs of cigarettes each day!

Much effort has been expended over the past four decades to mitigate the effects of fuelwood use among the nearly 3 billion people in the global South who still depend on biomass for their energy needs, mostly for cooking. But almost all of these efforts have been largely unsuccessful. The Global Alliance for Clean Cookstoves (GACC) has an ambitious plan to deploy 100 million High Efficiency Cookstoves (HECs) by the year 2020, but the plan has not yet been put into action due to technological and process hurdles. The Government of India has been trying to deploy HECs in the rural areas of India for the past two decades, but this intervention has been largely unsuccessful as well. At Climate Healers, we tried deploying solar cookstoves in the villages of Rajasthan and Orissa in 2010 and this was also unsuccessful[3]. Since then, we have been working with universities worldwide and mainly with Prof. Uday Kumar's team at the University of Iowa[4] and Prof. Bruce Litchfield's team at the University of Illinois[5] on stored energy solar cook stoves that can address the primary reasons for our unsuccessful deployment in 2010. But progress on these projects has been slow due to technical difficulties under our low-cost constraint. Meanwhile, the carbon offset mechanisms that Climate Healers planned to use for funding the deployment of these stored energy solar cookstoves have become mired in controversy and are largely defunct. Therefore, as of late 2014, we were open to consider a new course to get over these considerable procedural and technological hurdles.

## 8.1 Understanding the Problem

Over the past six years, Climate Healers' collaboration with the University of Iowa on the Winterim[6] program has been a tremendous boon to our healing efforts. The Winterim program, the brainchild of Prof. Rajagopal Rangaswamy, connects students and faculty from the University of Iowa to social projects in India so that the students experience fieldwork first hand. In our site in Rajasthan, India, the students spend three weeks over winter break working with Climate Healers and the Foundation for Ecological Security on the cookstove problem. A faculty member accompanies the students as the Winterim project earns the students 3 credits as well. While we get the benefit of a fresh set of eyes considering the problem and helping with the experiments, the students experience village life in India, usually for the first time ever!

In December 2014, it was as if the stars all aligned and everything came together to make a significant dent on the cookstove problem. Not only did we have a great group of enthusiastic Winterim students - as always - but we also had a multi-disciplinary team of faculty members from the University of Iowa accompanying them. The students were a diverse bunch, with liberal arts, engineering and health majors: Amanda Dolan, Jennifer Lilly, Sophie Mallaro, Michael Werner, AJ Walker, Naveen Ninan, Tim Wichert, Rohit Banda, Aditya Chahande, Julia Julstrom-Agoyo, Amber Mahoney, Rachel Maggi and Samantha Shannon. The faculty members were funded by a seed grant from the Center for Global and Regional Environmental Research (CGRER) at the University of Iowa to conduct research on the cookstove problem. They were Prof. Paul Greenough, a historian, Prof. Jerry Anthony, an architect, Prof. Marc Linderman, a geographer, Prof. Meena Khandelwal, Prof. Matthew Hill and Misha Quill, anthropologists and Prof. UdayKumar, a mechanical engineer. In addition, Michele Del Viscio, a mechanical engineer from Italy, volunteered for Climate Healers during that winter, to represent the perspectives of industry[7]!

When such a multi-disciplinary team was assembled in the villages of Rajasthan, India, all intent on solving the problem, good things happened! At first, we were focused on just testing two representative HECs in the villages to understand why their uptake was so poor in India. These were the best-rated HECs on the market and they cost $20-30, but even the villagers who could afford motorbikes were not using them. The uptake of HECs has been very poor in most places around the world. In India, less than

4% of the wood burning stoves are HECs[8]! This was a puzzle that we needed to first crack before we could attempt to solve the problem.

When compared to traditional "chulhas" (mud and brick stoves), HECs have several advantages. They have been designed scientifically with the best application of fluid mechanics and combustion principles to maximize the efficiency of the stove. In laboratory settings, these HECs do deliver the promised increased efficiency of up to 100% over traditional chulhas. We had to find out why real world performance did not match laboratory results.

The team conducted careful observations of the cooking process as the women of Karech and Gogunda tried to use the two HECs, labelled A and B. Through the gracious assistance of these women and the interpreters who helped us communicate with them, the main reasons for the poor uptake of these stoves in the villages of Rajasthan became quite clear. They are the following:

1. The commercial HECs don't accommodate the wide variety of wood fuel types that are available in Rajasthan. For instance, the HECs can't accept large pieces of wood without having them split lengthwise, which is very difficult for the women to do. When they face such a hurdle, the women tend to abandon these HECs since their chulhas have no such size limitation.

2. HEC Stove A heated the clay "tawa" (a vessel for cooking flatbreads called "rotis") too much in the center and not enough at the edges with the result that the women had to constantly rotate the rotis, especially the corn rotis, in order to cook their meal. Perhaps as a result, Stove A was not nearly as efficient in its use of firewood for cooking as advertised.

3. The mouth of HEC Stove B was too large to fit the clay tawas used in Rajasthan, with the result that we had to jerry-rig a grill to hold the clay tawa in place. Perhaps as a result, much of the advertised efficiency of HEC Stove B could not be obtained as well.

4. The women identified some safety issues with the HECs. The metal sides of the HECs got hot as the cooking progressed and the women expressed concern that their children could burn themselves if they touched the sides. In contrast, the traditional chulhas were made of clay and were naturally insulated.

5. The HECs exhibited visible deterioration in the three-week duration of the testing. The women expressed concern that these metal stoves may not last long and would need to be replaced as opposed to the traditional chulhas which last a couple of years and can be rebuilt with local material.

6. The women typically slow burn a large log in their traditional chulha to provide heating for their homes during the winter. The HECs could not accommodate such a large log and therefore could not be used for home heating.

7. Though there were some savings in firewood use with the HECs, the women estimated the stoves were worth as little as one-fifth the actual retail price of the HECs. Even then, it appeared doubtful that the women would actually pay that reduced amount to acquire such HEC stoves.

Traditional chulhas vary in size and shape to accommodate the different types of cooking vessels and foods cooked in them across the world. Our experience in Rajasthan showed that a single HEC stove couldn't possibly replace all these traditional stoves. Rather, significant fuelwood reductions can only be achieved with locally customizable solutions in different parts of the world. However, our tests with the HECs did confirm that they could reduce wood use significantly compared to the traditional chulhas. The low smoke effluence and main reductions in firewood use is due to the engineered airflow from below the fuel source in the HECs. Because it lacks such engineered airflow, the traditional chulha tends to accumulate embers that pile up and emit soot as they burn inefficiently due to a lack of oxygen. Overall, we measured anywhere from 30%-40% reductions in wood use with the HECs, when compared to the traditional chulha.

## 8.2 Stumbling on a Breakthrough

To experiment with the traditional chulha in a controlled setting, the team decided to build a makeshift chulha in the backyard of our hotel in Udaipur, Rajasthan. Though our makeshift chulha was going to be made with bricks and would not have mud caked around it as the traditional version in the villages, we felt that having a controlled setting would be useful for making comparisons. Once that chulha was built, I was tasked with starting the fire.

Perhaps the ground under the chulha was wet, or perhaps the firewood was wet, but for the life of me, I couldn't start the fire no matter how I arranged the sticks! Then I turned to Michele and asked him to light the fire.

Michele is one of the most practical engineers that I have ever encountered in my life. Throughout that Winterim, I was constantly amazed at how he could take whatever was lying around and use it to solve any knotty problem that we faced, whether it was HEC-B's mouth being too big, or whether we couldn't latch a gate in the village. But I was pleased to see that initially, even Michele couldn't start the fire in that makeshift chulha! Then, true to his reputation, Michele just walked around the backyard, picked out a grate from one of the unused HECs, stuck it in the chulha and lit the fire. Now the fire started almost instantly!

We eventually abandoned this experimental setup, since we were able to get the same user to work with all different stove configurations to cook their standard meals. But the fact that Michele used a metal grate to start the fire in that chulha stuck in my head. This was the first "Aha" moment.

Then, as were compiling a list of the problems that the women were experiencing with the HECs, we were struck by the fact that the traditional chulha had none of those problems. The only problem in the traditional chulha was that the airflow worsened over time due to the embers piling up. To address the airflow issue, we were bouncing around various engineering solutions - such as creating a grate on the floor of the chulha with a chamber underneath that would collect the ash, which we imagined connecting to an air duct from outside the house. Then the anthropologist among us, Matt, put his foot down firmly. He said that we must not even think about changing the traditional chulha in any way! The best we can do is to devise an accessory that the women can take away at any point to return to their old way of doing things.

That was the second "Aha" moment! That's when the idea of adding a grate to the traditional chulha flashed. A traditional stove burns with good thermal efficiency at the start of the cooking session, but the efficiency deteriorates over time. Typically, this efficiency starts at 15 percent and reduces down to 5%. However the grate that Michele had used to start the fire in the makeshift chulha was clearly not going to be of much use. It was designed for one of the HECs and its openings were large enough to allow the embers to fall down and pile up underneath it. The HEC itself had a secondary grate to drain the ash in the form of holes on its bottom plate, in addition to this coarse grate. Therefore, we fashioned a special raised metal grate that had holes to drain the ash and made sure the holes were too small for the embers to fall through. We had the design fabricated in the marketplace in Udaipur, took it to Lassibai's home in Karech and asked her to test it out in her traditional chulha.

Lassibai examined the grate and she was clearly not impressed with it. But she was game and inserted it in the chulha before she started her cooking session. Typically, when she starts, Lassibai brings a stack of wood from her pile in the back and keeps it near her. Based on her years of experience, she brings precisely the amount of wood she needs for her meal, almost to the stick. But this time, after finishing her cooking, she looked back and was amazed that more than half the stack that she had brought was still lying unused! So were we amazed, as well!

In fact, for the same meal, Lassibai had used 2.9 kgs of wood in the traditional chulha without the grate, 1.8 kgs with HECs A and B, and 1.1 kgs in the traditional chulha with the grate. That simple grate had caused a 60% reduction in wood use in the traditional chulha itself! I couldn't believe my eyes, but since we were at the end of the Winterim, we couldn't test the grate much further during that trip. Therefore, I made arrangements with my brother-in-law, Vasanth Kukillaya, who's a metal working expert, to engineer a robust version of the grate and have it tested at an official cookstove-testing center that is certified by the Government of India. We called this grate, the Mewar Angithi (MA), since this Angithi (a Hindi word for grate) was discovered in the Mewar district of Rajasthan.

It took about 3 months for the process to complete since we had to do paper work for the Cookstove Testing Center, schedule a date, and then Vasanth had to travel to Udaipur with two samples of the device to get it tested. After the first day of testing, Vasanth called me to say that Prof. Panwar and Prof. Rathore from the Maharana Pratap University affiliated with the Cookstove Testing Center were very animated at the end of the test and wanted to repeat it again the next day. Once the test was repeated, the Center released the official results: 63% reduction in wood use and 89% reduction in particulate matter (PM 2.5 μm) emissions for the same delivered energy! As soon as I got those official results, we set the ball rolling on field deployments and extensive testing in real world settings.

## 8.3 The Field Deployments

### The Setting: Rural Village in Rajasthan, India

There's no sign of the sun yet, but the sound of Kamlabai, the wife and mother of our host family, slipping on her shoes means it's not far from peeking over the horizon. Just like yesterday and tomorrow and virtually every day, she'll bring an armful of wood into the dark kitchen, start a fire, and prepare her rotis, a round flat bread made with water and corn or wheat flour. She'll prepare and cook each roti, one at a time, on a clay tawa over the flame in her traditional chulha. There is a certain charm in the routine: the rhythmic rolling and patting of fresh dough into consistently perfect rotis; the gradual awakening of the kids in the one-room house and of the livestock in the yard; the reliable illumination of the landscape complimenting this chore.

In summer 2015, Kayley Lain, an engineering graduate student at the University of Iowa, Nidhi Baid, an Environmental Engineering student at the State University of New York, our son, Sushil Rao and I spent time in five Rajasthani villages to arrange the distribution of 1000 MAs and test the performance of the device in the field. Our local NGO partner, Foundation for Ecological Security (FES), was responsible for much of the distribution process. Instead of mild steel that was used in the tested prototype, we used high carbon stainless steel (SS304H) for the MA to guard against possible corrosion from the chemicals in the wood. To reduce the cost of the device, we removed the bottom plate from the design. Testing of the MA revealed an average of 33% reduction in wood usage in seven households. Particulate matter (PM) reductions as high as 51% were observed in one household, with an average reduction of 33%.

Due to the high variability in smoke production observed in the field, we conducted additional lab-field hybrid tests at the University of Iowa. For these tests, we built a traditional chulha from bricks and cement in an 8'x8' outdoor tent, and heated a tawa to 350-400$^{\circ}$C, measuring wood usage and ambient PM. These tests are more repeatable than actual field tests – distractions to the cook, house design variations such as placement of windows, and variations in wood moisture are eliminated – but they are more realistic than the lab tests, allowing us to better predict field outcomes and potential design improvements. With this setup, we can collect more data in less time than we could with field tests, without leaving Iowa City. In these tests, we measured a 31% reduction in large (~10 μm) PM, which is in line with field data.

A follow-up field visit was conducted six months after families had received their MAs, to get feedback from users and inspect the MAs. The enthusiasm for the devices took us by surprise. Cooks said they don't cry from smoke anymore while they are cooking. They can make their meals faster and use less wood. None of the cooks we observed were ever excited about removing an MA so that we could record control data. A couple of them even tried to sneak it back into the chulha before starting the fire.

In a sample of 80 households in Rajasthan who received MAs, 71% of them used their MA regularly with no issues. None of these MAs had suffered any measurable weight loss or significant damage. Some women reported that they do not collect wood as many times a week as they did before they received an MA. Reasons for not using MAs included insufficient information upon receipt of the device and extra small chulha openings that could not accommodate the MA as supplied. Of the people who were able to use the MA, none reported any inconvenience induced by the device.

### The Setting: Orphanage in Kitui County, Kenya

The school children in Nyumbani Village have just finished their lunch in the canteen. Githeri, a maize and bean stew cooked over a fire, is served for lunch every day. Most of the children play football barefoot in the afternoon sun, but one 7-year-old boy, aptly nicknamed 'Little Engineer,' builds a toy truck from salvaged trash instead. He secretly sells these masterpieces to his classmates and uses the profits to buy sweets. Little Engineer lives with his susu ("grandmother") and nine other orphans in a house made of mud and concrete in Cluster 3. There are four houses in each cluster and 26 clusters in the village. Each house in the village is mostly the same as Little Engineer's – ten orphans and a susu or emau ("grandfather"). Besides time spent at school and on his budding toy car business, Little Engineer will spend one to two hours a day collecting firewood for one of his siblings to cook morning and evening meals for the family, which requires three to four hours a day.

In Nyumbani, Fabio Parigi and Michele Del Viscio found a wider range of stoves than in Rajasthan. The efficiencies of these stoves varied significantly. Using the MA in the most efficient of their stoves (which resemble the U-shaped chulhas from Rajasthan) increased the thermal efficiency by 25%. Replacing less efficient stoves (which consist of three rocks placed in a triangular arrangement) with the most efficient configuration and implementing MAs increased the thermal efficiency by 78%. These improvements translate into an estimated 7500 km$^2$ of local forest saved each year in Kenya alone.

## 8.4 The Future Opportunities

The MA has potential to spread organically by simply sharing the idea. The design is simple and flexible enough to be manufactured and distributed in localities around the world, which can also provide economic opportunities in small communities. Fabio and Michele have already sparked MA manufacturing at a school in Nyumbani. Students were able to make their own MAs with tools available to them in the village.

While measurements of emissions, both gaseous and particulate, have been performed with encouraging results, more particulate matter data is desired in field settings. Emissions are inherently difficult to characterize in the field because of the number of variables involved. The airflow and ventilation patterns in a house, the placement of recording instruments, the efficiency of the cook, moisture in the wood and several other factors can cause dramatic variations in emissions readings between households and over time. Also, emissions spike and drop regularly, with or without the MA. For these reasons, averages over numerous tests must be calculated. The value of these numbers relies on the number of data points used to calculate them.

Following more data on the field performance of the MA, we hope to characterize the impacts on the users of the MA and on their communities. Cooking is one aspect of life in the village, and it has complex interactions with other aspects of life. For example, livestock grazing and export of resources are additional causes of deforestation in many villages. With the time a woman can save on wood collection and cooking, she might increase the number of animals she raises, or she might have the opportunity to go to school. While most would consider the latter purely an improvement in quality of life, the former may cause more damage to the forest than it prevents.

The Mewar Angithi stove insert is affordable, effective, non-intrusive, and simple to make and distribute. It has the potential to reduce smoke-related diseases and illnesses in over 40% of the world's population and help protect forests that support people and local ecosystems. We aim to deploy more samples in India and Africa and conduct more extensive field-testing.

# 9. Towards Moral Singularity

*"Start by doing what's necessary, then what's possible, and suddenly you are doing the impossible"* - St. Francis of Assissi.

At present, we are being bombarded with apocalyptic stories predicting doom and gloom based on the linear extrapolation of our recent history. But we forget that our past is littered with nonlinear transformations that completely changed the linear course of history. For instance, I recall that in 1995 Newsweek ran an article asking if the Internet was going to go anywhere[1]? It was an extremely discouraging article for someone who was deep in the trenches of the internet revolution at that time. But ten years later, I overheard someone say that he couldn't live without the Internet!

Almost everyone agrees that the current global human population cannot be sustained forever. Biomass considerations alone preclude that. The current human biomass of 500 MT is 2.5 times larger than the biomass of all megafauna that the planet can be expected to sustain on a long term basis[2]. Therefore, a significant reduction in human population is called for, in the long term.

Given that such a population reduction is in order, a linear extrapolation into the future does indeed look apocalyptic in the current socioeconomic system, based on consumption as an organizing value. It's as if the Caterpillar, knowing that it has to become a Butterfly in order to reach sustainability, is imagining that a suitable liposuction followed by the grafting of some lightweight plastic wings might do the job, without truly undergoing a metamorphosis in the chrysalis. Such linear thinking explains the global inaction on climate change, biodiversity loss and desertification, the three major environmental problems that the UN Rio Summit had identified in 1992. As Elliott Sperber wrote recently[3],

"Instead of regarding the inability to act on climate change as a result of inertia or incompetence, perhaps we should begin to

regard it as willful. After all, who now sincerely doubts that
pollution and greenhouse gases create the conditions that
produce the ecological calamities that largely harm the poor?
And how can we overlook the related fact that the owners of
the world have a substantial incentive in ridding the planet of
the billions of people whose existence alone threatens their
property and privileges? Indeed, allowing climate change to
kill the poor would not only be more convenient than policing,
fighting, locking out and locking up billions; by claiming that
it's inevitable, the owners of the world can watch the ecological
holocaust unfold with a relatively good conscience. When one
considers this, along with the fact that the affluent classes
dictate social policy as well as the regulation of the pollutants
responsible for the climate calamities bombarding the (mostly)
poor, we may begin to see that the failure to halt the
proliferation of notoriously toxic gases is comparable to a type
of passive chemical warfare. Isn't that what it amounts to? And,
relevantly, there is a World War II precedent for just this type
of inaction as well. While the Red Army was losing millions in
their march toward Berlin, the US intentionally delayed
invading Europe in order to allow the Nazis to further weaken
the USSR, which the US, Britain, and others regarded as a
threat to their property (and the rule of money) ever since the
October Revolution."

The people of the global South are at the receiving end of this
strategy of inaction since poor countries are expected to be most
impacted by climate change[4]. Besides, if the world has reached a
point where people are dying in the billions, we can be sure that
global ecosystems would have collapsed as well. Therefore, this
strategy of inaction doesn't bode well for all Life on Earth, not just
for the downtrodden global South.

## 9.1 The Vegan Metamorphosis

The rise of Veganism is a swift kick in the rear to such linear
thinking. We know that the current socioeconomic system, based
on mindless consumption, is incompatible with Veganism. When
taken through its logical progression, Veganism necessitates

conscious simplicity since any unnecessary consumption uses natural resources that impacts animals somewhere. This is why going vegan is a process that doesn't stop with our dinner plates. This is also why "vegan consumerism" is an oxymoron and why it frightens the elites in the current system that more and more people are going vegan. Hence the widespread *Cowspiracy* that has infected elites and institutions, especially in the developed world. However, since no one can force people to consume animal products, the vegan metamorphosis is inexorable and just as in Nature, the Caterpillar has no choice but to become a Butterfly. As of 2015, 36% of Americans were actively experimenting with a plant-based vegan lifestyle[5]!

In this book, I have laid the case that the enormous waste and excess found in the animal agriculture industry can be used to devise a compassionate solution to our environmental predicament, which is considerate of all Life, not just humans. As ecosystems recover and human societies restructure around compassion, not consumption, as an organizing value, the demands on Nature will subside. As a globally equitable human society evolves, our descendants will naturally reduce their population over time and return towards the necessary balance with Nature. This is the true metamorphosis as the Caterpillar evolves into the Butterfly.

Buckminster Fuller once said[6],

"You never change things by fighting the existing reality. To change something, build a new model that makes the existing model obsolete."

In the self-improvement book, *Way of the Peaceful Warrior*, the main character, Socrates, also advised against fighting the old system head on[7]:

"The secret of change is focus all of your energies, not on fighting the old, but on building the new."

But in order not to fight the old, the old must not fight the new either. Therefore, it is important to build the new in a manner that complements the old and in harmony with it. Just as no one truly wants planetary catastrophes looming in our future as our current socioeconomic system continues to destroy the planet, no one wants the current socioeconomic system to collapse and cause global chaos either. That is precisely what world leaders are trying to avoid, but in their typically secretive, ham-handed fashion. Even the youth were not taken in by the charade that was the UN Paris Accord of 2015. Youth artists created toilet paper rolls with the text of the Paris Accord written on them and had those rolls smuggled into the restroom stalls at the UN COP-21 meeting in Paris!

Unfortunately, world leaders are forced to engage in this Kabuki theater. In a brief address to the press recently, President Obama described his job as follows[8]:

> "Being President is a hard job... Whoever's standing where I'm standing right now has the nuclear codes with him, and can order 21 year-olds into a firefight, and has to make sure that the banking system doesn't collapse and is often responsible for not just the United States of America but 20 other countries who are having big problems and are falling apart and are looking for us to do something."

Notice how stabilizing the banking system was high on his task list! Since the financial crisis of 2007-8, the big banks have become even bigger and almost universally disliked, because they are "too big to fail." They have successfully socialized their risks and privatized their profits. In fact, the total annual profits of the four largest banks in the US is about equal to their total federal subsidies, meaning that these banks are truly insolvent[9]. But the current debt-based financial system has become so complex, with unfettered trades in derivative instruments exceeding the Gross Domestic Product (GDP) of many large nations on a daily basis, that it might well be impossible to tinker with this financial system without triggering a collapse. As the current socioeconomic system chugs along destroying the planet, we just have a few years to work out its replacement. It is now in all our best interests, young and old, rich and poor, corporate and academic, to collaborate and build the infrastructure for a parallel new system that complements the old and can grow from it.

Today the mainstream media belittles Veganism as a fringe movement. I predict that ten years from now, Veganism will be mainstream and the average person will wonder aloud how we could have ever committed such atrocities against the animals and the planet!

I look forward to overhearing that conversation in ten years time.

## 9.2 Learning from Experience

System change is not a new imperative. People have been designing and implementing novel socioeconomic systems for decades, especially since the 1960s. Examples abound of intentional communities where people have redesigned their lives and their interactions with the environment[10]. But in a scenario reminiscent of what had been happening with respect to the "cookstove problem," none of these experiments have truly caught on. Most of them have obvious drawbacks that make it unlikely that they will be widely adopted.

First of all, there are very few intentional communities that follow a vegan lifestyle. If a community still does not understand the impact of an animal exploiting lifestyle, then I'm truly wary of what else are they failing to comprehend in their journey towards sustainability. Further, in my years of experience designing complex systems in the computer chip industry, I learned a simple truth:

> If you are not measuring and verifying any part of your design, then you can be absolutely sure that some aspect of it is broken!

Say what you will of our current Caterpillar system, but it does measure everything that matters to it. What matters to it is the monetization of Nature. The Caterpillar system turns water, life and even air into money and it is ruthlessly efficient at it. But as we implement prototype versions of Butterfly systems, we need to be equally diligent about measuring what matters to us, for example:

1. What is the total ecological footprint of the community?
2. Is the per capita ecological footprint compatible with the Half Earth solution?
3. Is there equality of opportunity for everyone in the community?
4. How much carbon is being sequestered through the regeneration of Life?

Another lesson that I learnt from my experience with the cookstove problem was the need to experiment in a safe space. In addressing that problem, what began as a solar cooker project ended up in the Mewar Angithi deployment! For the Mewar Angithi solution to emerge, we had to keep trying things out while being open to the idea that what we were doing wasn't going to be acceptable to the users! Likewise, we need to architect the Butterfly systems to be flexible and adaptive, where people of diverse backgrounds can come together and try things out and see what works in a safe, but real world setting.

The final lesson is that in the Butterfly system, it's a good idea to default to the exact opposite of what has become normal in the current Caterpillar system. For instance, in the current system, the corporations and the government are watching our every move on the Internet, while they guard almost everything they do as trade secrets or national security secrets. Therefore, the default in the current system is individual transparency and institutional privacy.

In the Butterfly system, it is a good idea to start with the exact opposite: the default should be individual privacy and institutional transparency. Every individual should be able to communicate anonymously and express freely. Everything that institutions do should be open-source and transparent. People need to be able to audit institutional books on an instant basis.

Above all, we need to have fun while working out the nuances of the Butterfly system over the next few years! After all, this is the greatest transformation in the history of human civilization and we're so lucky to be a part of it!

## 9.3 The Sacred Lifeline Project

Location, location, location!

Where should we set up the laboratories for the Butterfly system and showcase prototype implementations? At a side meeting of a large Interfaith gathering in Paris prior to COP-21, it was felt that we should target prominent sacred sites for such projects. Approximately, 100 million people travel to sacred sites around the world each year. Therefore, if these intentional communities are showcased in sacred sites, then it becomes easier to spread awareness of the emergence of the Butterfly.

The Sacred Lifeline project envisions a network of radically inclusive, sustainable, off-grid, zero-waste communities, modeling and exemplifying a compassionate, vegan lifestyle, while advancing the sustainability goals of the parent towns and cities. These communities will be located in well-recognized sacred sites around the world to provide opportunities for a steady flow of pilgrims and visitors to experience such a lifestyle during their visits. Roughly half the community would be mentors/educators and permanent residents while the other half would consist of short-term visitors, or longer-term college students working on sustainability projects. The communities will develop and adopt open-source software tools and technologies to ensure that their ecological footprint does not exceed half the Earth if the whole world were to live that way. The open-source software tools and technologies developed in the Sacred Lifeline project will then inspire and enable these pilgrims and visitors to continue their lifestyle experience remotely. The goal is to spread this view of how to operationalize sacredness to every part of the Earth until we realize a radically inclusive, equitable human society that is in harmony with a thriving natural world.

Living arrangements in the Sacred Lifeline project would be designed to facilitate individual privacy while connecting with

Nature and creating community around food production, preparation, sharing and enjoyment. We will implement the Sacred Lifeline project through partnerships with universities, academic institutions, NGOs, businesses and local markets, thereby connecting those who are transitioning towards a sustainable, compassionate lifestyle with suppliers of triple green products which are kind to humans (toxin-free), kind to the planet (pollution-free) and kind to animals (cruelty-free).

The Sacred Lifeline project will also help to close feedback loops in the current system, for example, through recycling waste water, composting food waste and agro-ecological farming. It will also provide affordable housing and healthy, organic, plant-based food preparation and delivery. There will be an Inter-Spiritual All-Faiths temple in each project location to signify the unity of all faith and wisdom traditions, including secular humanism and indigenous wisdom traditions, in this endeavor. The World Council of Religious Leaders is actively participating in the Sacred Lifeline project to bring the faith community onboard.

Each Sacred Lifeline centre will also offer specialized courses in meditation, yoga and other techniques to build resiliency in these troubled, transitional times. Revenues generated at each Sacred Lifeline centre would be used to fund re-wilding projects through the Earth Restoration Corps. The goal is to quickly realize the enormous carbon sequestration potential of regenerating native forests as the whole world goes vegan.

We are planning the inaugural Sacred Lifeline project in Crestone, Colorado, under the auspices of the Manitou Foundation[11], and in alignment with the enduring vision of Hanne Strong and her husband, the late Maurice Strong. To date, the Manitou Foundation has granted over two thousand acres of land to various spiritual, educational and environmental groups in Crestone and created a close-knit community of practitioners in the various faith and wisdom traditions of the world. In this "refuge for world truths" in the splendor of the Sangre de Cristo mountains, there are already over 25 different spiritual centers, including Buddhists, Shintos, Hindus, Sufis and Carmelite nuns. Not only is Crestone recognized as a spiritually uplifting place worldwide, but it is also a sacred place for the indigenous communities of North America. Therefore, it is an ideal place for a pioneering model community of the Sacred Lifeline project.

The Chrysalis Center at Crestone of the Sacred Lifeline Project will include a Vegan Experience Pavilion to connect Veganism and Ahimsa or non-violence in thought, word, and deed, as an essential response to climate change, biodiversity loss, desertification, toxic pollution and other environmental crises as well as our health, ethical and spiritual crises. We intend this Chrysalis Centre at Crestone to become a pilgrimage site for the global vegan community. The prime location of the Chrysalis Center presents greater visibility for Crestone in the global community.

With regard to local benefits, the surplus plant-based foods grown and prepared at the Chrysalis Center would be made available to the larger Crestone community and in return, the food waste from the Crestone community would be composted on the site in order to close the nutrient loop and replenish the soil.

## 9.4 The Moral Singularity

The search for spiritual awakening, the search for environmental sustainability and the search for social justice are all part of the same search for moral singularity, a state of being where we routinely experience the ultimate happiness that is already within us. This is the true pursuit of happiness, but to reach that state, we need to abandon the Cartesian viewpoint that had driven the Caterpillar:

I think. Therefore I am.

Such a formulation puts the monkey mind as the central actor in our lives and enhances the human ego. In the Butterfly, we reverse this perspective into:

I am. Therefore I act.

Life is about action and action driven by our true intuitive self is fundamentally compassionate. The thinking mind, like the forearm, is a limb that must function automatically in the background and not be out, front and center.

As individuals or as a species, it is only when we realize the ego's utter insignificance that we truly find that ultimate happiness. Yes, we are each the whole universe, but only when we become nothing inside. We are each powerful beyond measure, but only when we know that we are mere puppets.

For individuals, the Buddha's prescription for attaining Nirvana, the ultimate happiness or transcendental bliss, is deceptively simple. The Buddha articulated this in the Four Noble Truths[12]:

1. The world is full of suffering.
2. The root of suffering is attachments.
3. The cessation of suffering is through dropping attachments.
4. The liberation from suffering is through the Eightfold Noble path.

The Buddha laid out the Eightfold Noble path as a set of moral precepts:

1. Right View
2. Right Intention
3. Right Speech
4. Right Action
5. Right Livelihood
6. Right Effort
7. Right Mindfulness
8. Right Concentration

Yet over the next two millennia, Buddhism nearly died out in India, the land of the Buddha's birth. There was such a flowering of intellectual and philosophical discourse in India in the centuries following the Buddha's birth that it led to the assimilation of Buddhist teachings within the Yoga and Vedanta traditions of mainstream Hinduism.

Yoga is union and that union is of the body with the Universe, of the mind with Universal Consciousness and of the spirit with Brahman or God. It leads to the same ultimate happiness, the transcendental bliss of Nirvana, that the Yogis call Samadhi. The Yoga masters recognized that we are incapable of following moral precepts if we are physically ill conditioned, or if our mind is beset with fear, guilt, shame, grief, untruths, delusions as well as attachments. They devised systematic techniques for overcoming these physical and mental barriers through postures (*Asanas*), chanting (*Mantras*), controlled breathing (*Pranayama*) as well as meditation (*Dhyana*). While the Buddha recommended just observing our breath, the Yoga masters taught how to modify our breath consciously in pranayama. While the Buddha taught Vipassana or insight meditation that trained practitioners to drop their attachments from the outside in, the Yoga masters taught Chakra meditation that trained practitioners to let go of their fear, guilt, shame, grief, untruths, delusions and attachments, in that order, from the inside out. While in the Vipassana approach, our subconscious fears, guilt, shame, grief, etc., surface over time in an uncontrolled fashion, the Yoga practitioner can surface them and let them go consciously. This wealth of knowledge is available to us as we transform our global industrial civilization into its steady state, Butterfly version.

Thus far, the socioeconomic system of the global industrial civilization had kept us all rooted in fear and scarcity, the base chakra, which made the pursuit of happiness a total farce. The emerging new socioeconomic system will be rooted in compassion and abundance so that we can begin the pursuit of happiness in earnest. We will need to systematically let go of our guilt for having polluted the Earth so much, our shame for having exploited Nature so ruthlessly, our grief for all the biodiversity that we have destroyed and for all the animals that we have caused to suffer, our lies of exceptionalism that we had told ourselves, and our delusions of separation from Nature. Only then will we be in a position to let go our worldly attachments and become an enlightened member of the community of Life on Earth. Only then will we experience the transcendent happiness that is worth pursuing.

In a popular Conservation International video, the actress, Julia Roberts, intones[13]:

"Some call me Nature. Others call me Mother Nature. I've been here for over 4.5 billion years, 22,500 times longer than you. I don't really need people, but people need me. Yes, your future depends on me. When I thrive, you thrive. When I falter, you falter or worse. But I've been here for eons. I have fed species greater than you. And I have starved species greater than you. My oceans, my soil, my flowing streams, my forests, they all can take you. Or leave you. How you choose to live each day, whether you regard or disregard me doesn't really matter to me, one way or the other. Your actions will determine your fate, not mine. I am Nature. I will go on. I am prepared to evolve. Are you?"

Evolution is not a spectator sport. In an AhimsaCoin-like economy, when we all live as Pablo Picasso envisioned,

"The meaning of Life is to find your gift. The purpose of Life is to give it away,"

then humanity would have evolved as a species and the Butterfly would emerge. That is the journey towards moral singularity that is worth traveling.

It is my privilege to travel with you on this journey, dear reader! To my granddaughter, Kimaya, as Robert Frost wrote,

"I have promises to keep.
And miles to go before I sleep,
Miles to go before I sleep."

Welcome to the Vegan metamorphosis!

# Acknowledgments

*"The thing that brings about integration is love--- It's only love that sees and nothing else,"* Jiddu Krishnamurti.

Any book comes to fruition with the help of numerous reviewers, editors, publishers as well as the authors, and it is customary to acknowledge them. I'm especially grateful to Swami Sarvapriyananda for the lucid lectures on Vedanta that he has posted on Youtube and to Jayana Clerk for many stimulating conversations as the book took shape. The reviewers were amazing as they helped hone the narrative to its present form. I'm deeply indebted to Gillian Goslinga, Sai Praneeth, Meena Khandelwal, Jayana Clerk, Elemer Magaziner, Keith Akers, Lama Dawa, Sudha Fatima, Jon Thompson, Juan Ahonen-Jover, Deb Ozarko, Marty Landa, Pash Galbavy, Richard Pauli, Melvin Taylor, Richard Starling, Ezra Silk, Gani Ganapathi, H-S. Udaykumar, Kevin Greathouse, Mike Roddy, Linda Sills, Thomas Kailath, Roxy Chappell, Arun Sharma, Pamela Gale-Malhotra, Sudha Kukillaya, Umesh Rao, Sally Bingham, Rita Bruckstein and Freddy Bruckstein for plodding through the early drafts and sending much needed feedback. Gopi Govada's editing ironed out the grammatical errors and tensing issues in the draft.

This book contains extensive quotes from public sources as the story of humanity is integrated in the words of so many of us. Though most of these are involuntary contributors to the book, it is my privilege to acknowledge them all in the list below:

| | |
|---|---|
| Ahonen-Jover, Juan | Anderson, Kevin |
| Akers, Keith | Andubai of Peepulsarai |
| Akkineni, Amala | Anhang, Jeff |
| Alley, Richard | Annidevi of Kyarakhet |
| Alonso, Matthew | Anthony, Jerry |
| Amiyabai of Sakaria | Aurobindo, Sri |
| Andersen, Kip | |

# Bibliography

**Preface**

[1] Vipassana, which means to see things as they are, is one of India's most ancient techniques of meditation. For more information, please see, e.g., http://www.dhamma.org

[2] *A Study in Scarlet* is Sir Arthur Conan Doyle's first novel featuring Sherlock Holmes, perhaps the most famous fictional detective of all time. It can be found online at http://www.online-literature.com/doyle/study_scarlet/.

[3] The global temperature increase of 1°C has resulted in a 4% increase in moisture content of the atmosphere. This has caused extreme weather events such as the Chennai floods to become more frequent. Please see, e.g., https://en.wikipedia.org/wiki/2015_South_Indian_floods for more details.

[4] Prof. Donald Knuth uttered these words at the 2014 Kailath Lecture at Stanford University on May 7, 2014. The full lecture can be viewed online here: http://kailathlecture.stanford.edu/2014KailathLecture.html

[5] The segment aired in December 2005 on Link TV: http://www.linktv.org

[6] Rao, Sailesh, *Carbon Dharma: The Occupation of Butterflies*, Climate Healers, ISBN-13: 9781467928458, Oct. 2011. http://www.carbondharma.org

[7] McLaren, Brian D., *Everything Must Change: Jesus, Global Crisis and a Revolution of Hope*, Thomas Nelson, ISBN-13: 9780849901836, Oct. 2007, http://amzn.to/1NDFdi7

[8] The Plant-based Nutrition course is a joint offering of Cornell University and the T. Colin Campbell Center for Nutrition Studies. Details can be found at http://nutritionstudies.org/courses/plant-based-nutrition/.

[9] The Foundation for Ecological Security (FES) is a Non-Governmental Organization (NGO) that began as the National Tree Growers Cooperative Federation under the auspices of the National Dairy Development Board, Anand, Gujarat, India in 1988. More on FES at http://www.fes.org.in

[10] SAI Sanctuary was founded by Pamela and Dr. Anil K. Malhotra in 1991 in the Kodagu district of Karnataka, India. More on SAI Sanctuary at http://saisanctuary.com/

[11] The Climate Reality Project is the non-profit founded by former Vice President Al Gore to spread global awareness on climate change. More at https://www.climaterealityproject.org/

[12] The movement for American Indian rights aims to reverse policies that suppress the spiritual traditions of the indigenous communities of North America and help revive them. More, e.g., at http://aimovement.org/

[13] Ian Morris, *Why the West Rules for Now: The Patterns of History and What they Reveal About the Future*, Picador, Oct 2011, ISBN-13: 9780312611699. http://amzn.to/240JWMD

[14] The Booklist review was accessed on the Amazon web site at www.amazon.com/Why-West-Rules-Now-Patterns/dp/0312611692

[15] The Trans-Pacific Partnership (TPP) is a trade agreement between the 12 nations of the Pacific Rim that was signed on Feb. 12, 2016. It has not yet been ratified by the nations and therefore, has not entered into force yet. https://en.wikipedia.org/wiki/Trans-Pacific_Partnership

[16] The Rumi quote in question: "The wound is the place where the light enters you,"

https://www.goodreads.com/quotes/103315-the-wound-is-the-place
-where-the-light-enters-you

## 1. Our Stories Are Failing Us

[1] Jalal ad-Din Muhammad Rumi is one of my favorite Sufi mystics. A compilation of his quotes can be found here: http://www.goodreads.com/author/quotes/875661.Jalaluddin_Rumi

[2] Christopher Moore, *Practical DemonKeeping*, William Morrow, May 2004, ISBN-13: 978-0060735425, http://amzn.to/1WHMAHO

[3] Please see Harari, Yuval Noah, *Sapiens: A Brief History of Humankind*, Harper, Feb 2015, ISBN-13: 978-0062316097, http://amzn.to/1WHMfF4

[4] The scientific method as we know it today is attributed to Francis Bacon, *Novum Organum*, published in 1620, though variants had appeared much earlier.

[5] These top ten problems first appeared in Smalley, Richard, *"Top Ten Problems of Humanity for Next 50 Years,"* Energy and Nanotechnology Conference, Rice University, 2003.

[6] These statistics were reported by Gro Brundtland in her Stanford lecture accessed here: https://www.youtube.com/watch?v=NSC97yEjJDU

[7] These statistical findings were reported in the 2014 Living Planet report of the World Wildlife Fund. It can be accessed here: http://www.worldwildlife.org/pages/living-planet-report-2014

[8] Population and Consumption figures can be obtained from the UN Human Development Reports, e.g., http://hdr.undp.org/en/content/human-development-report-2014

[9] This passage is from the Oct 2014 column, http://www.theguardian.com/environment/georgemonbiot/2014/oct/01/george-monbiot-war-on-the-living-world-wildlife

[10] Please see Drew Hansen's column: http://onforb.es/25d0t5X

[11] Gilding, Paul, *The Great Disruption: Why the Climate Crisis Will Bring On the End of Shopping and the Birth of a New World,* Bloomsbury Press, March 2011, ISBN-13: 978-1608192236, http://amzn.to/2bIh7k6

[12] The quote is taken from Paul Gilding's TED talk here: http://www.ted.com/talks/paul_gilding_the_earth_is_full

[13] More than 3 billion people, almost half the population of the world, live on less than $2 per day. http://www.globalissues.org/article/26/poverty-facts-and-stats

[14] Please see Prof. Jonathan Turley's analysis here: http://bit.ly/1fpWHg7

[15] Moynihan, R., Heath, I., Henry, D., "Selling Sickness: The Pharmaceutical Industry and Disease Mongering," BMJ. 2002 Apr 13; 324(7342): 886–891. http://bit.ly/2bCr2gL

[16] The parable of the broken window, introduced by Frederic Bastiat in 1850, is commonly understood to be a fallacy, but it gets applied widely in an oligarchic setting. More on the parable here: https://en.wikipedia.org/wiki/Parable_of_the_broken_window

[17] Moore, Gordon, IEEE 1975 Speech, http://bit.ly/1OYSDPs

[18] Intel's former chief architect, Bob Colwell, made this prediction in a recent speech. Mr. Colwell now heads DARPA's MicroSystems Technology Office. Please see, e.g., http://bit.ly/1niX9iK

[19] John Oliver's net neutrality segment can be found here: https://www.youtube.com/watch?v=fpbOEoRrHyU

[20] 1000BASE-T has now become a commodity item that can be purchased as a core for integration in larger devices: http://bit.ly/1sEiKYV

[21] 10GBASE-T shipment numbers for 2011 can be found here. http://bit.ly/1YHtwaw . The poor uptake of 10GBASE-T on

existing cabling has prompted the development of 2.5GBASE-T
and 5GBASE-T variations under the NGBASE-T alliance.

[22] Sipe, Richard, *A Secret World: Sexuality And The Search For Celibacy,* Routledge, Jun 2014, ISBN-13: 978-1138004740

[23] The UN Convention on Biological Diversity can be found
online at http://www.cbd.int , the UN Convention to Combat
Desertification at http://www.unccd.int and the UN Framework
Convention on Climate Change at http://www.unfccc.int .

[24] 49% of the people in the US have some form of anxiety
disorder, depression or substance abuse issues. All of them are
taking some form of chemical supplements to address these mental
problems.

https://www.psychologytoday.com/blog/anxiety-files/200804/how-big-problem-is-anxiety

The average high school kid in the US today has the same level of
anxiety as the average psychiatric ward patient in the 1950s.

[25] Please see this article and the references cited therein:
https://www.thefix.com/content/wall-street-addiction-finance-cocaine-meltdown7456

[26] Please see video on dioxins in the food supply and references
cited: http://nutritionfacts.org/video/dioxins-in-the-food-supply/

[27] The popular narrative is reflected in this New York Times
opinion piece:
http://www.nytimes.com/2015/12/16/opinion/paris-climate-accord-is-a-big-big-deal.html

[28] This quote is taken from an article on the Stanford MAHB web
site: http://stanford.io/1Tr704i

[29] Based on the Global Footprint Network's assessment that the
Earth Overshoot Day is August 8, 2016. This means that 1.6 Earths
are needed to support human demand today.
http://www.overshootday.org/

[30] Based on a Production and Materials report from the UNEP's International Panel on Sustainable Resource Management. Please see article and references cited therein: http://www.theguardian.com/environment/2010/jun/02/un-report-m eat-free-diet

[31] Dr. Kevin Anderson stated this in a discussion taped during COP-21 in Paris: https://www.youtube.com/watch?v=svlU6p0gHgo

[32] This characterization is cited in http://www.yaleclimateconnections.org/2015/12/jim-hansen-pans-c op21-baloney/

[33] Charles Eisenstein made this statement in a Youtube video segment here: https://www.youtube.com/watch?v=XSetJdaJm28

[34] This declaration was signed during COP-17 in Durban, South Africa, as reported here: http://www.climatehealers.org/blog/2015/8/9/update-on-cop-17-fro m-durban-south-africa

[35] This view was popularized in Daniel Quinn's book, *Ishmael: An Adventure of the Mind and Spirit*, Bantam, May 1995, ISBN-13: 978-0553375404, http://amzn.to/1NDF5z8

[36] The estimate of the maximum population that the Earth can support depends upon the lifestyle of the storyteller.

[37] A video of Ms. Sering making this statement can be found in http://www.democracynow.org/2013/11/20/filipino_climate_chief_i t_feel_like

[38] https://en.wikipedia.org/wiki/Rapture

[39] George Carlin made this statement in a 1992 HBO special https://www.youtube.com/watch?v=EjmtSkl53h4

[40] A systems model predicting this as one possible scenario: http://bit.ly/1gjyS3u

## 2. Separation is a Delusion

[1] It is customary for a parent to feel this way when his or her child is born, not when a grandchild is born. Please see the Prologue section in *Carbon Dharma: The Occupation of Butterflies*, http://www.carbondharma.org

[2] Artists have imagined fantastic planets, but not another that is also life-sustaining.

[3] The Bertrand Russell quote can be accessed here, https://www.goodreads.com/quotes/11219-the-greatest-challenge-to -any-thinker-is-stating-the-problem

[4] The quote can be found in Chapter 4 of Ophuls, William, *Immoderate Greatness: Why Civilizations Fail*, CreateSpace Independent Publishing, ISBN-13: 978-1479243143, Dec. 2012, http://amzn.to/1WHM8t8

[5] http://www.saisanctuary.org

[6] Diamandis, Peter H., Kotler, Steven, *Abundance: The Future Is Better Than You Think,* Free Press, Feb 2012, ISBN-13: 978-1451614213

[7] http://bit.ly/1smriwy

[8] A timeline on legal slavery can be found in http://www.pbs.org/wnet/slavery/timeline/1865.html

[9] https://en.wikipedia.org/wiki/British_Raj

[10] The last ethnographic display of humans in a zoo was in Belgium in 1958: http://www.bbc.com/news/magazine-16295827

[11] http://bit.ly/1SROoe0

[12] https://www.youtube.com/watch?v=-PgMETKrQD4

[13] Rao, Sailesh, *Carbon Dharma: The Occupation of Butterflies*, Climate Healers, ISBN-13: 9781467928458, Oct. 2011. http://www.carbondharma.org

[14] Anodea, Judith, *Waking the Global Heart*, Energy Psychology Press, ISBN-13: 978-0972002899, Jan 2010. http://amzn.to/1NLCuTF

[15] The reference is to Donald Trump's campaign to "*Make America Great Again*," while purging it of Muslims, Mexicans and other minorities. Of late, such nationalistic campaigns have proliferated in the Western world.

[16] A Global Witness report found that 116 environmental activists were murdered in the year 2014, more than double the number of journalists murdered. Most of the environmental activists were indigenous people protesting mining and hydro projects. https://www.globalwitness.org/en/campaigns/environmental-activists/how-many-more/

Berta Caceras and Nelson Garcia were murdered within the span of two weeks in Honduras. http://www.telesurtv.net/english/news/Another-Member-of-Berta-Caceres-Group-Assassinated-in-Honduras-20160315-0049.html

[17] Please see, e.g., Will Potter, *Green is the New Red: An Insider's Account of a Social Movement Under Siege*, City Light Publishers, April 2011, ISBN-13: 978-0872865389, http://amzn.to/1s0UKi0

[18] Please see E. O. Wilson, *Half-Earth: Our Planet's Fight for Life*, Liveright, March 2016, ISBN-13: 978-1631490828, http://amzn.to/1Rfrzwb

[19] http://www.hinduclimatedeclaration2015.org/

[20] http://bit.ly/1XDSxoB

[21] http://bit.ly/1WBlJvA

[22] http://bit.ly/1sEsIcR

[23] http://bit.ly/1W5CRep

[24] http://bit.ly/22hP6Vl

[25] http://bit.ly/23ZmPlv

[26] http://bit.ly/1XDTeOJ

[27] http://www.interfaithdeclaration.org/

[28] Leslie, Robert, *Yoga Psychology: The Science of the Inward Connection*, iUniverse, Apr 2006, ISBN-13: 978-0595393688, http://amzn.to/2c9IPHz

[29] The material in this section was framed through extensive discussions with the notable author, speaker, educator and my surrogate mother in Sedona, AZ, Jayana Clerk.

[30] Please see E. O. Wilson, *Half-Earth: Our Planet's Fight for Life*, Liveright, March 2016, ISBN-13: 978-1631490828. http://amzn.to/1Rfrzwb

[31] http://savitrithepoem.com/

[32] https://en.wikipedia.org/wiki/Savitri_and_Satyavan

[33] Please see Sri Aurobindo's commentary in http://savitrithepoem.com/the-poem.html

[34] Please check out the Forbes article and references therein: http://onforb.es/1sva0DM

[35] In a private conversation as reported in Swami Prabhavananda's book, *The Spiritual Heritage of India: A Clear Summary of Indian Philosophy and Religion* (1979).

## 3. Everything is Perfect

[1] Various English translations of this passage can be found in http://biblehub.com/job/42-2.htm .

[2] Attributes of Allah can be found in verses 59:22-24 of the Holy Quran, for example: "He is God; there is no god but He. He is the Knower of the unseen and the visible; He is the All-Merciful, the All-Compassionate. He is God; there is no god but He. He is the King, the All-Holy, the All-Peace, the Guardian of the Faith, the All-Preserver, the All-Mighty, the All-Compeller, the All-Sublime. Glory be to God, above that they associate! He is God, the Creator, the Maker, the Shaper. To Him belong the Names Most Beautiful. All that is in the heavens and the earth magnifies Him; He is the Almighty, the All-Wise" See, e.g., http://www.sultan.org/articles/god.html .

[3] An excellent exposition of the definition of Brahman in the Taittiriya Upanishad can be found in Swami Sarvapriyananda's lecture: https://www.youtube.com/watch?v=Ftn4zCnheBk .

[4] Pope Francis' Encyclical, the *Laudato Si* can be downloaded here: http://laudatosi.com/watch .

[5] The Islamic Declaration on Climate Change can be found online here: http://islamicclimatedeclaration.org/islamic-declaration-on-global-climate-change/ .

[6] The text of the Hindu Declaration on Climate Change can be found online here: http://www.hinduclimatedeclaration2015.org/english .

[7] Robert Oppenheimer uses this quote in the 1965 BBC movie made about the atomic bomb: http://www.atomicarchive.com/Movies/Movie8.shtml .

[8] Various English translations of this passage can be found here: http://biblehub.com/isaiah/43-7.htm .

[9] This quote can be found in the book Harari, Y-N, *Sapiens: A Brief History of Humankind*, Harper, Feb 2015, ISBN-13: 978-0062316097, http://amzn.to/1WHMfF4

[10] Conversation as reported in http://cnet.co/1VdQVBt .

[11] The quest to extend the human lifespan is well funded: http://bit.ly/1YIyO5G .

[12] Hindu worldly pursuits are known as Purusharthas: http://bit.ly/1rYQj7m .

[13] The text of the declaration can be found here: http://www.climatehealers.org/blog/2015/8/9/update-on-cop-17-from-durban-south-africa .

[14] The Dalai Lama was quoting from his book, *Beyond Religion: Ethics for a Whole World*, Rider, Jan 2013, ISBN-13: 978-1846043109.

[15] See, e.g., https://en.wikipedia.org/wiki/Ahimsa .

[16] Anthony DeMello, *Awareness: The Perils and Opportunities of Reality*, Image Publishers, June 1990, ISBN-13: 978-0385249379, http://amzn.to/25eUxt6

[17] This quote is taken from Richard Feynman's 1966 address to science teachers: http://www.feynman.com/science/what-is-science/ .

[18] Examples abound of such egregious actions: http://bit.ly/1XnNTuw .

[19] We can monitor forest loss real time at http://www.globalforestwatch.org/ .

[20] Perkins, John, *Confessions of an Economic Hit Man*, Plume, Dec 2005, ISBN-13: 978-0452287082.

[21] Of the prison population of 2.3 million people in the US, nearly 1 million are African Americans: http://www.naacp.org/pages/criminal-justice-fact-sheet .

[22] Please see Fraser and Freedman's article in http://www.salon.com/2012/04/19/21st_century_chain_gangs/ .

[23] Taken from Prof. Eben Moglen's lecture at Columbia University in November, 2013: http://snowdenandthefuture.info/PartIII.html .

[24] The Lacey Act is featured in the video on why you shouldn't talk to the police: https://www.youtube.com/watch?v=6wXkI4t7nuc .

[25] Hartney, Christopher and Vuong, Linh, "Created Equal: Racial and Ethnic Disparities in the US Criminal Justice System," National Council on Crime and Delinquency Report, March 2009. http://www.nccdglobal.org/sites/default/files/publication_pdf/create d-equal.pdf .

[26] https://en.wikipedia.org/wiki/Neo-Nazism

[27] https://en.wikipedia.org/wiki/Caste_system_in_India .

[28] https://en.wikipedia.org/wiki/Ageism

[29] Quote taken from Tuttle, Will, ed., *Circles of Compassion: Essays Connecting Issues of Justice*, Vegan Publishers, Jan 2015, ISBN-13: 978-1940184067, http://amzn.to/1TsYKAO

[30] See news article in http://dailym.ai/1TuGjgg .

[31] See news article in http://bit.ly/1OQsGYL .

[32] https://en.wikipedia.org/wiki/Homophobia

[33] Quote taken from the video: http://bit.ly/1JNlrZx

[34] https://en.wikipedia.org/wiki/Carnism

[35] Singer, Isaac Bashevis, *Enemies, a Love Story,* Farrar, Strauss and Giroux, Jan 1997, ISBN-13: 978-0374515225, http://amzn.to/1WIAY81

[36] Please see article http://bit.ly/1nWABFe

[37] https://en.wikipedia.org/wiki/Animal_Enterprise_Terrorism_Act

[38] https://en.wikipedia.org/wiki/Ag-gag

[39] https://en.wikipedia.org/wiki/Second_Congo_War

[40] Lochbaum et al., *Fukushima: The Story of a Nuclear Disaster,* The New Press, Feb 2014, ISBN-13: 978-1595589088, http://amzn.to/1U6ll3h

[41] The quote is taken from the interview here: https://www.wagingpeace.org/general-lee-butler/ .

[42] http://www.mysterium.com/amnh.html. See also, Barnosky, A. D., et al., "Approaching a State Shift in Earth's Biosphere," Nature, 486, pp. 52–58, 07 June 2012, doi:10.1038/nature11018, , http://go.nature.com/1JF0gw5

[43] http://time.com/4340269/antarctica-climate-change-ice-melt/

[44] Notz, D., "The future of ice sheets and sea ice: Between reversible retreat and unstoppable loss," Proceedings of the National Academy of Sciences, vol. 106 no. 49, pp. 20590–20595, doi: 10.1073/pnas.0902356106, http://bit.ly/2bTsNDX

[45] Smil, Vaclav, *Harvesting the Biosphere: What We Have Taken from Nature,* MIT Press, Dec 2012, ISBN-13: 978-0262018562, http://amzn.to/20jbSdA

[46] https://news.stanford.edu/2005/06/14/jobs-061505/

[47] http://www.quoteworld.org/quotes/5566

[48] http://bit.ly/1TrWolG

[49] Anthony DeMello, *Awareness: The Perils and Opportunities of Reality,* Image Publishers, June 1990, ISBN-13: 978-0385249379, http://amzn.to/25eUxt6

## 4. Who Are We?

[1] http://www.yesmagazine.org/pdf/kortennewstory.pdf

[2] This passage is quoted by Swami Sarvapriyananda in https://www.youtube.com/watch?v=Ftn4zCnheBk

[3] https://en.wikipedia.org/wiki/Dvaita

[4] https://en.wikipedia.org/wiki/Vishishtadvaita

[5] https://en.wikipedia.org/wiki/Advaita_Vedanta

[6] The relevant experiments are described in http://bit.ly/1TrZSov .

[7] http://james-mcwilliams.com/?p=5524

[8] https://en.wikipedia.org/wiki/Accipitridae

[9] http://nyti.ms/1XnUfdj

[10] https://en.wikipedia.org/wiki/Bloodhound

[11] https://en.wikipedia.org/wiki/Alpine_ibex

[12] https://en.wikipedia.org/wiki/Peregrine_falcon

[13] https://en.wikipedia.org/wiki/Mitochondrial_Eve

[14] http://bit.ly/1qz1Fxm

[15] Human partnership with dogs formed an evolutionary advantage for humans as explored in Shipman, Pat, *The Invaders: How Humans and Their Dogs Drove Neanderthals to Extinction,* Belknap Press, March 2015, ISBN-13: 978-0674736764, http://amzn.to/1WIBAKO

[16] Research on Dolphin language is documented at http://www.speakdolphin.com/home.cfm .

[17] Slobodchikoff, Con, *Chasing Doctor Dolittle: Learning the Language of Animals,* St. Martin's Press, Nov 2012, ISBN-13: 978-0312611798, http://amzn.to/1YLFf86

[18] Quote found in http://www.yesmagazine.org/pdf/kortennewstory.pdf

[19] http://www.pigeon.psy.tufts.edu/psych26/darwin1.htm

[20]
http://fcmconference.org/img/CambridgeDeclarationOnConsciousn
ess.pdf

[21] Fossat et al., "Anxiety-like behavior in crayfish is controlled by
serotonin," Science, vol. 344, issue 6189, pp. 1293-1297, Jun 2014,
http://bit.ly/2crXH8S

[22] Culum Brown, et al., *Fish Cognition and Behavior*,
Wiley-Blackwell, Aug 2011, ISBN-13: 978-1444332216,
http://amzn.to/1YLFjEJ

[23]
http://www.teachingsofthebuddha.com/see-yourself-in-others.htm

[24] See, e.g., https://en.wikipedia.org/wiki/Ahimsa .

[25] https://www.youtube.com/watch?v=c_DV-Ul9AM

[26] https://www.utm.edu/staff/jfieser/class/160/9-animals.htm

[27] Quote taken from Tuttle, Will, ed., *Circles of Compassion:
Essays Connecting Issues of Justice*, Vegan Publishers, Jan 2015,
ISBN-13: 978-1940184067. http://amzn.to/1TsYKAO

[28] Dr. King channeled these Theodore Parker's words concisely on
several occasions, including at the Wesleyan University
Baccalaureate ceremony in 1964.
http://quoteinvestigator.com/2012/11/15/arc-of-universe/

[29] Please see http://www.ncbi.nlm.nih.gov/pubmed/19562864/

[30] The quote is taken from
http://nutritionfacts.org/2011/09/08/how-much-pus-is-there-in-milk

[31] Quote found in
https://www.bragg.com/healthinfo/hereshealth.html

[32] https://www.climaterealityproject.org/

[33] http://www.climatehealers.org

[34] http://www.fes.org.in

[35]
http://time.com/3833931/india-beef-exports-rise-ban-buffalo-meat/

[36] https://www.youtube.com/watch?v=LU8DDYz68kM

[37] https://www.dhamma.org/

[38] Please see, e.g., http://bit.ly/ZZ3uH7

[39] Lisle, Doug and Goldhamer, Alan, *The Pleasure Trap: Mastering the Hidden Force That Undermines Health and Happiness*, Healthy Living Publications, Mar 2006, ISBN-13: 978-1570671975, http://amzn.to/1TyavHp

[40] https://en.wikipedia.org/wiki/Nature_deficit_disorder

[41] http://www.ncbi.nlm.nih.gov/pubmed/18607383/

[42] https://www.youtube.com/watch?v=a52vAx9HaCI

[43] Based on GDP growth projections by PriceWaterhouse Coopers with respect to 2015: http://pwc.to/1MEl8UJ

[44] Gilding, Paul, *The Great Disruption: Why the Climate Crisis Will Bring On the End of Shopping and the Birth of a New World*, Bloomsbury Press, March 2011, ISBN-13: 978-1608192236, http://amzn.to/1W8jqlf

[45] http://store.hartman-group.com/culture-of-millennials-2011/

[46] https://www.google.com/trends/explore#q=vegan

[47] Rao, Sailesh, *Carbon Dharma: The Occupation of Butterflies*, Climate Healers, ISBN-13: 9781467928458, Oct. 2011. http://www.carbondharma.org

**5. What is Our Relationship with the World?**

[1] The relevant experiments are described in http://bit.ly/1TrZSov .

[2] https://www.youtube.com/watch?v=MzW-r_vPf50

[3] This is the dialog in the Bhagavad Gita,
http://www.bhagavad-gita.org/index-english.html

[4] Prof. Seligman explains positive psychology here:
http://www.pursuit-of-happiness.org/history-of-happiness/martin-se
ligman-positive-psychology/

[5] Prof. Seligman talks about philanthropy vs entertainment here:
http://www.ted.com/talks/martin_seligman_on_the_state_of_psych
ology

[6] Here's a compilation of Swami Vivekananda's writings on
mental concentration:
http://www.swamivivekanandaquotes.org/2013/12/swami-vivekana
ndas-quotes-on_4.html

[7] Sri Ramakrishna Paramahansa on free will can be found here:
http://sfvedanta.org/monthly-reading/sri-ramakrishna-on-free-will/

[8] Sam Harris talks about compassion during his lecture on free
will: https://www.youtube.com/watch?v=pCofmZlC72g

[9] Tony Robbins describes his breakdown of needs in his TED talk:
https://www.ted.com/talks/tony_robbins_asks_why_we_do_what_
we_do

[10] George Monbiot reports on the connection between extinction
and human arrival in this column:
http://www.monbiot.com/2014/03/24/destroyer-of-worlds/

[11] Please read Prof. Will Steffen's assessment here:
http://www.isthishowyoufeel.com/this-is-how-scientists-feel.html#
will

[12] Please see, e.g.,
http://www.firstpeoples.org/who-are-indigenous-peoples/how-our-s
ocieties-work

[13] The PDF of an English translation can be found online here:
http://www.mkgandhi.org/ebks/hind_swaraj.pdf

[14] Klein Goldewijk, K., Beusen, A., Van Drecht, G. and De Vos, M., 2010: "The HYDE 3.1 spatially explicit database of human induced global land-use change over the past 12,000 years," Global Ecology and Biogeography, vol. 20, issue 1, pp. 73-86, doi: 10.1111/j.1466-8238.2010.00587.x

[15] https://en.wikipedia.org/wiki/Milankovitch_cycles

[16] https://en.wikipedia.org/wiki/Neolithic_Revolution

[17] McKibben, Bill, *Eaarth: Making a Life on a Tough New Planet*, St. Martin's Griffin, Mar 2011, ISBN-13: 978-0312541194, http://amzn.to/1XoQxQV

[18] Quote is taken from Bill Ruddiman's lecture abstract:
http://bit.ly/1NCCbul

[19] A calculation of the radiative forcing due to the Milankovitch cycles is found here:
http://www.skepticalscience.com/Milankovitch.html

[20] Here's a discussion on why two degrees is an crucial number:
http://www.pbs.org/newshour/bb/why-2-degrees-celsius-is-climate-changes-magic-number/

[21] The Global Footprint Network calculates an ecological overshoot day here: http://www.overshootday.org/

[22] A PDF scan of the original 1972 book, Limits to Growth, can be found here:
http://www.donellameadows.org/wp-content/userfiles/Limits-to-Growth-digital-scan-version.pdf

[23] Economists such as Julian Simon argued that human ingenuity will overcome any scarcities as it is the ultimate resource. His 1981 book "Ultimate Resource" is now in its second version: Simon, Julian, *The Ultimate Resource 2*, Princeton University Press, Oct 1996, ISBN-13: 978-0691042695, http://amzn.to/1sw7gpC

[24] Graham Turner's article in the Guardian can be accessed here: http://bit.ly/1tVayzy

[25] J. Rockstrom, et al., "Planetary Boundaries: Exploring the Safe Operating Space for Humanity," vol 14, no 2, art 32, 2009. http://www.ecologyandsociety.org/vol14/iss2/art32/

[26] Hansen, J., et al, "Target atmospheric CO2: Where should humanity aim?," Open Atmos. Sci. J. (2008), vol. 2, pp. 217-231, 2008. http://arxiv.org/abs/0804.1126

[27] Annual mortality due to atmospheric pollution is broken down here: http://www.who.int/mediacentre/factsheets/fs313/en/

[28] Habitat loss is the primary cause of species extinctions: http://wwf.panda.org/about_our_earth/species/problems/habitat_los
s_degradation/

[29] Please use the flow diagram on Page 836 of the document: https://www.ipcc.ch/pdf/assessment-report/ar5/wg3/ipcc_wg3_ar5_
chapter11.pdf

[30] Barnosky, A., "Megafauna Biomass Tradeoff as a Driver of Quaternary and Future Extinctions," Proceedings of the National Academy of Sciences, vol. 105, suppl. 1, pp. 11543-11548, 2008, doi: 10.1073/pnas.0801918105. http://bit.ly/2bDbfZ0

[31] https://en.wikipedia.org/wiki/Bycatch

[32] http://www.takeextinctionoffyourplate.com/

[33] Goodland, R and Anhang, J, "Livestock and Climate Change," Worldwatch Institute report, 2009. http://bit.ly/TYv58d

[34] Herrero et al, "Livestock's Long Shadow," FAO report, 2006. http://bit.ly/1uTv8Tl

[35] A good explanation of the IPCC Tier 1, 2 and 3 reporting mechanisms can be found in http://bit.ly/1YJJIbg

[36] A good discussion of the 20 year vs 100 year time window can be found in http://bit.ly/1J8CUO2

[37] Rao, Sailesh, *Carbon Dharma: The Occupation of Butterflies*, Climate Healers, ISBN-13: 9781467928458, Oct. 2011. http://www.carbondharma.org

[38] Herrero, et al. "Livestock and Greenhouse Gas Emissions: The Importance of Getting the Numbers Right," Volumes 166-167, Pages 779–782, June 2011, DOI: http://dx.doi.org/10.1016/j.anifeedsci.2011.04.083

[39] Goodland, R and Anhang, J, "Livestock and greenhouse gas emissions: The importance of getting the numbers right, by Herrero et al. [Anim. Feed Sci. Technol. 166–167, 779–782]," Volume 172, Issues 3-4, Pages 252–256, March 2012, DOI: http://dx.doi.org/10.1016/j.anifeedsci.2011.12.028

[40] There is a notation on the Animal Feed Science and Technology journal page that Herrero et al declined to continue the debate. http://dx.doi.org/10.1016/j.anifeedsci.2011.12.028 . Please see explanation at http://bit.ly/240nYcD

[41] Gerber, P.J., et al, Tackling climate change through livestock – A global assessment of emissions and mitigation opportunities, FAO, 2013, http://www.fao.org/3/i3437e.pdf

[42] Barnosky, A., "Megafauna Biomass Tradeoff as a Driver of Quaternary and Future Extinctions," Proceedings of the National Academy of Sciences, vol. 105, suppl. 1, pp. 11543-11548, 2008, doi: 10.1073/pnas.0801918105, http://bit.ly/2bDbfZ0

[43] Chianese, et al, "Simulation of Carbon DiOxide Emissions from Dairy Farms to Assess Greenhouse Gas Reduction Strategies," Transactions of ASABE, vol 52, no 4, pp. 1301-1312, 2009. http://1.usa.gov/256NvDf

[44] Jacobson, M. F. "More and Cleaner Water." In Six Arguments for a Greener Diet: How a More Plant-based Diet Could save Your Health and the Environment. Washington, DC: Center for Science

in the Public Interest, 2006.
http://www.cspinet.org/EatingGreen/pdf/arguments4.pdf

[45] Allen, et al. "Fossil Fuels and Water Quality," Chapter 4, The
World's Water, Vol 7, 2012, http://bit.ly/1FouO2h

[46] A 2014 satellite study confirmed that livestock produced more
methane than fossil fuel sources in the US: http://bit.ly/1uMxHDj

[47] Szidat et al, "Contributions of fossil fuel, biomass-burning, and
biogenic emissions to carbonaceous aerosols in Zurich as traced by
14C", J. Geophysical Res., Volume 111, Issue D7, April 2006,
http://onlinelibrary.wiley.com/doi/10.1029/2005JD006590/full

[48] Prof. T. Colin Campbell estimates that 70% of the
pharmaceutical intake can be avoided if humans stop consuming
animal foods, http://www.forksoverknives.com/the-film/

[49] This estimate is based on the observation that 35% of the land
area of the planet is currently used to grow fodder to feed livestock.
In addition, half the biomass from cropland is consumed by
livestock as well. The argument being made is that if the biomass
from cropland is fed directly to humans, then the grazing land for
cattle will be freed up for ecosystem restoration. See IPCC AR5
WG3 Chapter 11, page 836:
https://www.ipcc.ch/pdf/assessment-report/ar5/wg3/ipcc_wg3_ar5_
chapter11.pdf

[50] This assessment is based on the observation that grain-feeding
of livestock is prevalent mainly in the affluent nations of the world.

[51] This is a lower bound since crops that don't use chemical
fertilizers, as grown in the developing world, are used for human
consumption directly.

[52] Rao, S., Jain, A. K., Shu, S., "The Lifestyle Carbon Dividend:
Assessment of the Carbon Sequestration Potential of Grasslands
and Pasturelands Reverted to Native Forests," AGU Fall Meeting,
Dec 2015,
https://agu.confex.com/agu/fm15/meetingapp.cgi/Paper/67429

[53] Hansen et al, "Doubling Down on Our Faustian Bargain," Huffingtonpost, Mar 2013, http://huff.to/1OED5l0

[54] The quote is taken from James McWilliams blog, http://james-mcwilliams.com/?p=341

[55] As characterized by Philip Wollen in http://bit.ly/1JNlrZx

[56] This viewpoint was popularized by Alan Savory in his TED talk, http://bit.ly/1kI51ft . However, this has been repeatedly debunked, e.g., at http://slate.me/1cbRZuJ

[57] A chicken grown in 1940 took 14 weeks to maturity and could fly. A chicken grown today takes 6 weeks to maturity and can barely walk. http://bit.ly/1eVWLk2

[58] http://www.cowspiracy.com/

[59]
http://www.goodreads.com/quotes/6908-the-time-is-always-right-to-do-the-right-thing

[60] http://www.cowspiracy.com/facts/

[61] From Kahneman, D, *Thinking Fast And Slow*, Farrar, Strauss and Giroux, Apr 2013, ISBN-13: 978-0374533557, http://amzn.to/IOzzcc

[62]
http://www.goodreads.com/quotes/369-a-human-being-is-a-part-of-the-whole-called

[63] http://movies.disney.com/cinderella

## 6. Why Are We Here?

[1] Berry, Thomas, *The Dream of the Earth*, Counterpoint, Oct 2006, ISBN-13: 978-1578051359, http://amzn.to/240x0WZ

[2] As popularized in the photograph snapped by Apollo 17 astronauts, http://go.nasa.gov/1N7w9hb

[3] Lovelock, J, *Gaia: A New Look at Life on Earth*, Oxford University Press, Nov 2000, ISBN-13: 978-0192862181, http://amzn.to/20h7coD

[4] Bill O'Reilly has opined on climate numerous times. Here's a clip where he articulates this position: http://mm4a.org/22izKzH

[5] An archetypal event is the so-called "Climate Gate" scandal, which was debunked subsequently: http://bit.ly/1OTAoM2

[6] Makarieva, A. M., et al, "Precipitation on land versus distance from the ocean: Evidence for a forest pump of atmospheric moisture," Ecol. Complexity, Volume 6, Issue 3, September 2009, Pages 302–307, doi:10.1016/j.ecocom.2008.11.004, 2009. http://bit.ly/240yl0a

[7] Alley, Richard, "The Biggest Control Knob," AGU Plenary lecture, 2012, https://www.youtube.com/watch?v=RffPSrRpq_g

[8] Dorn, Ronald, "Ants as a powerful biotic agent of olivine and plagioclase dissolution," Geology, 2014, doi:10.1130/G35825.1, http://bit.ly/256ZoJs

[9] Ward, R.C., "The Spirits Will Leave: Preventing the Desecration and Destruction of Native American Sacred Sites on Federal Land," Ecology Law Quarterly, vol 19, issue 4, art 4, Sep 1992. http://bit.ly/27JxxRN

[10] Page 137 in Berry, Thomas, *The Dream of the Earth*, Counterpoint, Oct 2006, ISBN-13: 978-1578051359, http://amzn.to/240x0WZ

[11] https://en.wikipedia.org/wiki/Extinction_event

[12] Raup, D.; Sepkoski Jr, J. "Mass extinctions in the marine fossil record". Science 215 (4539): 1501–1503. 1982. doi:10.1126/science.215.4539.1501 http://bit.ly/1XFYfqg

[13] https://en.wikipedia.org/wiki/Extinction_event

[14] Here is an article from 2003 on QQ47: http://bit.ly/25egxEF

[15] Here is an article from 2003 downgrading QQ47:
http://bit.ly/1svJaeE

[16] Campbell, et al, "The Impact Imperative: Laser Ablation for
Deflecting Asteroids, Meteoroids, and Comets from Impacting the
Earth," AIP Conf. Proc. 664, 509 (2003);
http://dx.doi.org/10.1063/1.1582138

[17] https://en.wikipedia.org/wiki/Biodiversity

[18] https://en.wikipedia.org/wiki/Galactic_year

[19] https://en.wikipedia.org/wiki/G-type_main-sequence_star

[20] Kasting et al, "Habitable Zones Around Main Sequence Stars,"
ICARUS 101, 108-128, 1993. http://bit.ly/1TpFbat

[21] Kopparapu, R, et al, "Habitable Zones Around Main-Sequence
Stars: New Estimates," Earth and Planetary Astrophysics, 2013,
Doi: 10.1088/0004-637X/765/2/131, http://arxiv.org/abs/1301.6674

[22] http://climate.nasa.gov/climate_resources/24/

[23]
https://www.goodreads.com/author/quotes/5387.Pierre_Teilhard_de
_Chardin

## 7. Everything Will Change

[1] https://www.youtube.com/watch?v=urQPraeeY0w

[2] https://en.wikipedia.org/wiki/The_Social_Contract

[3] The PDF of an English translation can be found online here:
http://www.mkgandhi.org/ebks/hind_swaraj.pdf

[4] http://bit.ly/1qzUMfp

[5] http://www.zeitgeistmovingforward.com/

[6] The quote is taken from his article on "The Insatiable God,"
http://www.monbiot.com/2014/11/18/the-insatiable-god/

[7] Paul Krugman wrote this in 2014 http://nyti.ms/1wu9C59

[8]
http://www.goodreads.com/quotes/643596-what-does-economic-gr
owth-actually-mean-it-means-more-consumption

[9] As quoted in this blog article: http://bit.ly/1t7mMkl

[10] As reported in the Guardian here: http://bit.ly/13uCrWh

[11] As quoted in http://bit.ly/Zc8x7j

[12] Erik Lindberg's article is found here:
http://www.resilience.org/stories/2014-11-26/six-myths-about-clim
ate-change-that-liberals-rarely-question

[13] A transcript of Severn Suzuki's speech can be found here:
http://www.americanrhetoric.com/speeches/severnsuzukiunearthsu
mmit.htm

[14] Severn Cullis-Suzuki's speech from 2012 can be found here:
http://www.unfoundation.org/news-and-media/press-releases/2012/
rioplussocial-six-minute-speech.html

[15] Please see the blog article here:
http://kevinanderson.info/blog/why-carbon-prices-cant-deliver-the-
2c-target/

[16] This statistic is found in the 30 year update to the Limits to
Growth, Meadows, et al, *Limits to Growth: The 30-Year Update*,
Chelsea Green publishing, 2004, ISBN-13: 978-1931498586,
http://amzn.to/1W7sVB6

[17] Google's relies on advertising revenues, unlike Wikipedia
which operates on donations. http://mklnd.com/1PRqf8E

[18] As quoted in
http://grist.org/food/vandana-shiva-so-right-and-yet-so-wrong/

[19] Ian Morris, *Why the West Rules for Now: The Patterns of History and What they Reveal About the Future*, Picador, Oct 2011, ISBN-13: 9780312611699. http://amzn.to/240JWMD

[20] Buckminster Fuller, *Utopia or Oblivion: The Prospects for Humanity*, Lars Muller publishers, 2008, ISBN-13: 978-3037781272, http://amzn.to/2578sxY

[21] As reported in http://nyti.ms/1t5z0Mx

[22] See fact sheet and references cited therein: http://www2.nami.org/factsheets/mentalillness_factsheet.pdf

[23] 49% of the people in the US have some form of anxiety disorder, depression or substance abuse issues: https://www.psychologytoday.com/blog/anxiety-files/200804/how-big-problem-is-anxiety

[24] http://nyti.ms/1oNpMaf

[25] Polling results compiled when Congressional approval ratings were at 9%: http://bit.ly/1qnwRwh

[26] Gilens and Page's study is summarized here: http://slate.me/1KHqTBT

[27] James D'Angelo's findings are analyzed here: http://ivn.us/2015/07/16/transparency-the-greatest-flaw-in-congress

[28] Quoted from this article: https://chomsky.info/20130604/

[29] Please see, e.g., http://bit.ly/1TpNLWG

[30] Please see, e.g., http://bit.ly/1U6hdQU

[31] Quote taken from Tuttle, Will, ed., *Circles of Compassion: Essays Connecting Issues of Justice*, Vegan Publishers, Jan 2015, ISBN-13: 978-1940184067. http://amzn.to/1TsYKAO

[32] Please see e.g.,
https://en.wikipedia.org/wiki/Declaration_on_the_Rights_of_Indig
enous_Peoples

[33] Also mentioned in
https://en.wikipedia.org/wiki/Declaration_on_the_Rights_of_Indig
enous_Peoples

[34] Pejorative college and professional sports team names raise
hackles in the indigenous communities:
http://www.huffingtonpost.com/2014/12/29/redskins-protest-home-
game_n_6390570.html

[35] Bruce Schneier wrote this after the Snowden revelations
http://www.cnn.com/2013/03/16/opinion/schneier-internet-surveilla
nce/

[36] Congress specifically exempted the oil and gas industry from
environmental laws and from disclosure requirements:
http://bit.ly/1U5nsHw

[37] http://www.c-span.org/video/?c4510300/rosa-brooks-testimony

[38] Prof. Joern Fischer's quote is taken from this video:
https://www.youtube.com/watch?v=e_UKw6Z6OP0

[39] This quote is taken from the 2014 G20 Communique:
http://1.usa.gov/1YLaXlQ

[40] http://www.pnas.org/content/106/8/2483.full

[41] Tuttle, Will, ed., *Circles of Compassion: Essays Connecting
Issues of Justice*, Vegan Publishers, Jan 2015, ISBN-13:
978-1940184067. http://amzn.to/1TsYKAO

[42] Holly Wilson in Pojman, P and Pojman, L, ed., *Food Ethics*,
Wadsworth publishing, 2011, ISBN-13: 978-1111772307,
http://amzn.to/1NDsnQE

[43] Allport, Gordon, *The Nature of Prejudice: 25th Anniversary Edition*, Basic books, 1979, ISBN-13: 978-0201001792, http://amzn.to/1VersaY

[44] David Hufton in Vanhoute, K and Lang, M, *Bulling and the Abuse of Power*, IDP Publishing, 2010, ISBN: 978-1-84888-045-0, http://bit.ly/1TxzyKL

[45] In Vanhoute, K and Lang, M, *Bullying and the Abuse of Power*, IDP Publishing, 2010, ISBN: 978-1-84888-045-0, http://bit.ly/1TxzyKL

[46] The ratio is derived from the 70 billion animals who are raised and slaughtered each year as opposed to the billion plus human victims, who are bullied in their lifetimes.

[47] Dr. Will Tuttle, *The World Peace Diet: Eating for Spiritual Health and Social Harmony*, Lantern Books, 2004, ISBN-13: 978-1590560839, http://amzn.to/1ONMp5r

[48] https://en.wikipedia.org/wiki/Speciesism

[49] http://veganfeministagitator.blogspot.com/

[50] Jeremy Rifkin discussed these three drivers in an interview with Een Vandaag entitled On Global Issues and Future of the Planet, https://www.youtube.com/watch?v=m9wM-p8wTq4

[51] http://stanford.io/17arxa8

[52] https://paulgilding.com/2014/03/19/carbon-crash-solar-dawn/

[53] https://www.youtube.com/watch?v=5yB3n9fu-rM

[54] http://www.cdc.gov/tobacco/data_statistics/sgr/history/

[55] http://www.tobaccoatlas.org/topic/cigarette-use-globally/

[56] Here's a list of 15 surprising things that contain animal products: http://read.bi/PXAdYQ. This is just the tip of the iceberg.

[57] The Network of Equality Giving was instrumental in targeting the dog-whistle politics of the Democratic establishment: http://www.equalitygiving.org

[58] https://en.wikipedia.org/wiki/Letter_from_Birmingham_Jail

[59] https://laudatosi.com/watch

[60] Pope Francis's meal selections were reported in http://fxn.ws/1Fzjwvz

[61] http://www.ncbi.nlm.nih.gov/pubmed/19562864

[62] Gandhi's speech is transcribed in its entirety here: http://www.ivu.org/news/evu/other/gandhi2.html

[63] This headline was taken from Time Magazine in Nov 2014, http://ti.me/27KEDp9 . The Gadhimai festival takes place every five years and the organizers have announced that it will be discontinued in the future: https://en.wikipedia.org/wiki/Gadhimai_festival

[64] http://www.animals24-7.org/2014/03/12/427/

[65] https://en.wikipedia.org/wiki/Drift_netting

[66] http://www.cowspiracy.com

[67] The article on Chris Hedges' transition: http://bit.ly/1ONeAin

[68] Johann Hari's article is here: http://www.thenation.com/article/wrong-kind-green-2/

[69] Here's the announcement of the partnership between Al Gore's Climate Reality project and Ben and Jerry's ice creams: http://bit.ly/241n3IW

[70] http://www.whatthehealthfilm.com/

[71] http://www.meatlessmonday.com/

[72] Bittman, Mark, *VB6: Eat Vegan Before 6:00 to Lose Weight and Restore Your Health . . . for Good*, Clarkson Potter, 2013, ISBN-13: 978-0385344746, http://amzn.to/1U5vYGv

[73] Here's an NPR segment on this issue: http://n.pr/1QN9qHA

[74] http://www.ucobserver.org/society/2016/05/vegans/

[75] https://en.wikipedia.org/wiki/Khadi

[76] http://bit.ly/1XoMSCo

[77] Brown, Rebecca, *Gandhi's Spinning Wheel and the Making of India (Routledge Studies in South Asian History)*, Routledge, 2010, ISBN-13: 978-0415494311, http://amzn.to/1XoN9Ft

[78] Please see, e.g., http://bit.ly/1PpQFuG

[79] http://www.mkgandhi.org/ebks/gandhijionkhadi.pdf

[80] http://www.bbc.com/news/uk-england-lancashire-15020097

[81] Statistics about Germany reported in http://bit.ly/1IMqh7Y

[82] http://sacred-economics.com/

[83] See discussion in http://bit.ly/257PeIE

[84] Quigley, Carroll, *Tragedy & Hope: A History of the World in Our Time*, GSG and Associates, 2004, ISBN-13: 978-0945001102, http://amzn.to/1zot71w

[85] Quote taken from Thomas Greco's presentation, "Money, Power, Democracy and War," http://slideplayer.com/slide/8007644/

[86] Quote taken from article: http://www.thenation.com/article/capitalism-vs-climate/

[87] https://www.youtube.com/watch?v=3YR4CseY9pk

[88] Here's a reasoned criticism of Klein's position: http://bit.ly/1TqyZ1Y

[89] https://www.youtube.com/watch?v=DyV0OfU3-FU

[90] https://en.wikipedia.org/wiki/Bitcoin

[91] Please see E. O. Wilson, *Half-Earth: Our Planet's Fight for Life*, Liveright, March 2016, ISBN-13: 978-1631490828. http://amzn.to/1Rfrzwb

[92] Quote is taken from http://www.zeitgeistmovingforward.com/

[93] See, e.g., http://inequality.org/inequality-health/

[94] This assumes a world population of 7.4 billion people and a planetary capacity of 15 Billion global hectares.

[95] A detailed comparison can be found here: http://bit.ly/1XoPQH8

[96] Wilson's law was first reported on the New York Times Dot Earth blog: http://nyti.ms/1OFzxiv

[97] E. O. Wilson, *Half-Earth: Our Planet's Fight for Life*, Liveright, March 2016, ISBN-13: 978-1631490828. http://amzn.to/1Rfrzwb

[98] http://nyti.ms/1LTxntX

[99] Monbiot promotes re-wilding here: http://bit.ly/1IcyF70

[100] Kolbert, Elizabeth, *The Sixth Extinction: An Unnatural History*, Henry Holt & Co, 2014, ISBN-13: 978-0805092998, http://amzn.to/1sw6xVq

**8. The Mewar Angithi:**

[1] Kaimowitz, D. "What Causes Tropical Deforestation?," Forestry Chronicle 78(3), 359–359 (2002).

[2] Grieshop, A.P., et al. "Health and Climate Benefits of Cookstove Replacement Options," Energy Policy 39(12), 7530–7542 (2011)

[3] http://www.climatehealers.org/solar-cook-stoves/

[4] http://bit.ly/25eVkub

[5] http://bit.ly/1XoRgl9

[6] http://bit.ly/1DksJPR

[7] http://www.thesolutionsjournal.com/node/237379

[8] Dalberg Global Development Advisors. Global Alliance for Clean Cookstoves: India Cookstoves and Fuels Market Assessment [online] (2013), http://bit.ly/1ZFisLf

**9. Towards Moral Singularity:**

[1] Originally entitled, "The Internet? Bah!," Newsweek has re-titled the article, "Why the Web Won't be Nirvana," and it can be found here: http://bit.ly/18Bpo4k

[2] Barnosky, A., "Megafauna Biomass Tradeoff as a Driver of Quaternary and Future Extinctions," Proceedings of the National Academy of Sciences, vol. 105, suppl. 1, pp. 11543-11548, 2008, doi: 10.1073/pnas.0801918105, http://bit.ly/2bDbfZ0

[3] Elliott Sperber, "Clean, Green, Class War: Bill McKibben's Shortsighted 'War on Climate Change'," Counterpunch, Aug. 22, 2016, http://bit.ly/2bFBzqu

[4] Hallegatte, S., et al., "Shock Waves: Managing the Impacts of Climate Change on Poverty," World Bank report, 2015, http://bit.ly/1MjUDl1

[5] This statistic was reported in the 2015 Special Diets Report of the Nutrition Business Journal here: http://newhope.com/2015-nbj-special-diets-report-and-webinar. This report is behind a pay wall, but an analysis explaining the statistic in question can be found here: http://bit.ly/1HLKZqT

[6] http://www.goodreads.com/quotes/13119-you-never-change-things-by-fighting-the-existing-reality-to

[7] Attributed to the character, Socrates, in Dan Millman, WAY OF THE PEACEFUL WARRIOR: A Book That Changes Lives, HJ

Kramer/New World Library, 2009, ISBN-13: 978-1932073256,
http://amzn.to/25804OS

[8] https://www.youtube.com/watch?v=oDmIdhcQ4Ew

[9]
https://www.warren.senate.gov/files/documents/Rigged_Justice_201
6.pdf

[10] http://gen.ecovillage.org/

[11] http://www.manitou.org/

[12] http://www.buddhaweb.org

[13] https://www.youtube.com/watch?v=WmVLcj-XKnM

# Index